Underwater Welding of Offshore Platforms and Pipelines

Proceedings of a Conference
November 5-6, 1980
New Orleans, Louisiana

Sponsored by
American Welding Society Education Department

Welding Technology Series
American Welding Society
2501 N.W. 7th Street, Miami, FL 33125

Library of Congress Number: 80-70761
International Standard Book Number: 0-87171-215-6

Printed in the United States of America

Contents

Underwater Welding of Offshore Platforms and Pipelines
Conference Program Advisory Board

Buddy Delaune, Jr.
Senior Engineer
Taylor Diving & Salvage Company
Belle Chasse, LA

Conrad J. Schoppe
Research Division Supervisor
Tennessee Gas Pipeline Company
Houston, TX

Yolanda Smith
Manager, Conferences and Seminars
American Welding Society
Miami, FL

Bruce J. Sylvester
President
Sylvester Underseas Inspection
Rockland, MA

James D. Theisen
Manager, Welding, Quality Control and Training
Brown & Root, WHMC
Houston, TX

Dwight L. Turner
Welding Engineer
J. Ray McDermott & Company, Inc.
New Orleans, LA

James L. Watts
Surveyor, Hull Technical Staff
American Bureau of Shipping
New Orleans, LA

Debrah Weir
Director, Education
American Welding Society
Miami, FL

Preface

This highly successful conference was designed to offer insight into emerging technology of great importance in future pipeline and offshore platform design and construction, including practical information on diving and equipment options, welding processes adaptable to the underwater environment, maintenance and repair operations, safety, and the latest AWS recommended practices for underwater welding.

AWS feels that the more than 250 welding engineers, designers, offshore maintenance personnel, QA and QC managers, inspectors, and NDT technicians who attended "Underwater Welding of Offshore Platforms and Pipelines," November 5-6, 1980 did so because of the planning of the Conference Program Advisory Board: P.T. Delaune, Senior Engineer, Taylor Diving and Salvage Co., Belle Chasse, LA; C.E. Grubbs, President, D & W Underwater Welding Services, Inc., Slaughter, LA; Conrad J. Schoppe, Research Div. Supervisor, Tennessee Gas Pipeline Co., Houston; Bruce J. Sylvester, President, Sylvester Underseas Inspection, Rockland, MA; James D. Theisen, Manager, Welding Quality Control and Training, Brown and Root, WHMC, Houston; Dwight L. Turner, Welding Engineer, J. Ray McDermott & Co., Inc., New Orleans, and James L. Watts, Surveyor, Hull Technical Staff. American Bureau of Shipping, New Orleans.

The two days of the conference focused on hard-hitting facts on the construction, maintenance, and repair of offshore platforms and pipelines and highlighted these objectives:

- what available equipment can upgrade underwater welding productive performance
- advantages (and disadvantages) of the submerged one atmosphere welding system
- how to logically schedule inspection, maintenance, and repair work on underwater structures
- newest underwater ultrasonic thickness gauging system
- proven procedures and methods to follow for a successful major underwater project
- superiority of GMAW–underwater
- critical factors affecting metallurgical and mechanical properties of underwater welds
- where OCS regulations are heading

- how simulation produced welding procedures that increased safety margins underwater
- how to adapt fitness-for-purpose as THE underwater welding criterion
- present limitations in saturation diving
- how to obtain quality welds through proper training
- when explosive welding is the proper alternative to conventional joining
- which equipment is the safest and most efficient
- how to minimize both inspection time and diving costs in repair or maintenance operations
- the newest in stereophotographic systems

An important objective was how AWS proposed Underwater Welding Specification would impact attendee operations. In fact, a goal of an AWS conference is to provide a forum for AWS Technical Committees to introduce latest standards, specifications, and codes. This opportunity has been used in many conferences in the past and was successfully handled at "Underwater Welding of Offshore Platforms and Pipelines" by AWS D3b Committee member E.A. Silva.

The international flavor of "Underwater Welding of Offshore Platforms and Pipelines" was represented by the line-up of speakers from not only the U.S. but from England and France. Attendees were attracted from the U.S., Canada, England, France, Mexico, Japan, and the Scandinavian countries. In fact, of the 7,000 conference or seminar attendees expected by AWS to participate in one or more of our programs, roughly 20 percent are targeted as overseas delegates.

AWS is indebted to the volunteers who contributed so much of their time and effort to create the successful "Underwater Welding of Offshore Platforms and Pipelines" – the speakers: H.C. Cotton, Welding Engineering Consultant, London, England; Michael C. Bateman-Cooke, Project Manager, Taylor Diving and Salvage Co., Belle Chasse, LA; Walter T. Bugno, Project Engineer, Standard Oil Co. of California, San Francisco; David Groves, Materials Engineering Manager, Brown and Root (U.K.) Ltd., London, England; Herman B. Smith, Head, Test and Development Branch, Welding Engineering Div., Norfolk Naval Shipyard, Chesapeake, VA; Edward M. Briggs, Director, Div. of Ocean Engineering and Robert E. Adler, Manager, Ocean Systems Section, Southwest Research Institute, San Antonio, TX; Gilbert Coriatt, Manager of Hyperbaric Welding, Comex Services, Marseille, France; Dr. David J. Leidel, Senior Research Engineer, Jet Research Center, Inc., Arlington, TX; Koichi Masubuchi, Professor, Massachusetts Institute of Technology; David B. Wyman, Senior Project Engineer, Diver Tools Div.,

Naval Coastal Systems Center, Panama City, FL; Thomas J. Dawson, Metallurgical Engineering Consultant, Naval Facilities Engineering Command, Alexandria, VA; E.A. Silva, Ocean Technology Program Director, Office of Naval Research, Dept. of the Navy; Richard J. Giangerelli, Structural Engineer, U.S. Geological Survey, Reston, VA; Bernard G. Sudreau, Project Manager, Pipeline Dept., Compagnie Francaise Des Petroles and J. Paul Gaudin, Ingenieur Manager, Hyperbaric Laboratory, Institut De Soudure, Paris, France; John W. Robinson, trial attorney, Gretna, LA; and Bruce J. Sylvester, Sylvester Underseas Inspection.

The AWS New Orleans Section was especially helpful in recruiting attendees to join AWS as Associate Members and we particularly thank Mr. Shelton L. Ritter of Braun Welding Supply for his untiring efforts.

Debrah Weir

The Challenge of Deep Sea Diving

Michael C. Bateman-Cooke
Taylor Diving and Salvage Company, Inc.

Introduction

Down through the centuries, technology has decreased step by step the limitations of diving, but probably more in the past decade than ever before. The original limitations were very basic things, such as how long could one hold one's breath, how deep could one breathe with a snorkel, and how long could one survive on the air contained in a diving bell. The first real breakthrough occurred around the turn of the 19th century with the invention of the air pump and the closed deep sea diving rig by August Siebe in 1837. Today, however, there are many methods of diving employed in offshore oil fields, depending upon the nature of work, the depth of water, and the required bottom time. This paper will deal basically with the rather recent and rapid progression of deep diving over the past decade and a half and its relationship to the advancement of hyperbaric welding.

General

Anyone who has participated in the performance of a deep water hyperbaric weld will certainly agree that the successful welded connection, repair, and attachment of two (2) pieces of steel in a dry environment is still one of the most ambitious and challenging projects with which the diving industry is confronted today.

It is not the intention of this paper to elaborate on the history of diving and its progression from shallow air diving through surface mixed gas and eventually to that mode known as saturation. However, the majority of hyperbaric welding being performed is by saturated diver/welders, and, for the benefit of those who are not familiar with the expression "saturation diving," a brief explanation is in order.

In conventional diving (that is, surface-to-surface without the use of a bell), the diver is decompressed during the ascent. This is accomplished by following a precalculated schedule that consists of a prescribed rate of ascent plus a number of stages or stops. This provides for the elimination of gases that have been forced into solution in the various body tissues by escaping without forming bubbles in the blood stream and thus causing decompression sickness or bends. The time of the dive is limited by what is considered a reasonable amount of decompression. In deep water, the dive must be relatively brief in order to avoid an unbearable decompression schedule. Employing the saturation diving technique, the divers are sealed in a deck-board chamber and pressurized close to the depth pressure at which they intend to work. A diving bell at the same pressure may be attached to or detached from the deck chamber so the divers can be lowered away under pressure to the job site and, upon completion of their mission, be returned to the deckboard chamber still under pressure. After 12 hours at a given pressure the body is said to be saturated; that is, it has absorbed or taken into solution the maximum amount of gas it is capable of containing at that pressure level. The main object of saturation diving is to eliminate decompression after each and every dive; thus, the length of time the diver may spend working at depth is limited only by his physical endurance. Finally, when the project is completed after a matter of days or weeks, the divers are then decompressed in the deckboard chamber by very slowly lowering the pressure while observing certain stops at predetermined levels as part of the schedule. This method necessarily requires that an artificial atmosphere be maintained in the bell and chamber and that a somewhat complicated environmental control system be employed. Temperature and humidity levels must be maintained. Carbon dioxide and other harmful gases must be removed, while oxygen has to be maintained at a constant level.

While all helium and oxygen (HeO_2) total saturation equipment is built around the same basic theory, it varies in physical configuration from one company to another. Even the nomenclature varies. For instance, the diving bell itself, which we call a submersible diving chamber (abbreviated SDC), is referred to by many other companies as a personnel transfer capsule (abbreviated PTC). One of the main differences is in the variety of ways the SDC is attached to the deckboard decompression chamber (DDC). There are various davit-like arrangements for handling the SDC on the surface and mating it to the DDC. Some SDCs referred to as the side-entry type have a transfer door in the side that flanges to a transfer tunnel in the end of the DDC, while others mate to a transfer hatch on top of a DDC and the personnel transfer is made through the hatch in the bottom of the SDC. Most of our saturation equipment is designed in this manner. A structural steel base frame, which extends over the side of the support platform, is fitted with an A-frame on the outboard end. The SDC is lowered and raised from the A-frame. When on the surface in position for transfer, the SDC is held firmly in the guides of the A-frame and locked in a fixed position. The DDC is mounted on skid runners that match channels of the base frame. By means of a hydraulic cylinder, the DDC is slid outboard until the flange of the transfer tunnel matches with the side-entry flange of the SDC.

While many of the SDCs are cylindrical in shape with either hemispherical or elliptical heads, others are spherical and built as dual pressure vessels.

Total saturation HeO_2 diving is the ultimate technique in the industry. However, it is seldom used at depths shallower than 150 feet, except in the case of hyperbaric welding. The reason for this is that most people consider saturation diving not economically feasible in shallow water or in deep water where only an occasional dive has to be made. A surface-to-surface compressed air diving system has a value of from approximately $20,000.00 to $35,000.00; surface-to-surface HeO_2 diving systems range from approximately $50,000.00 to $100,000.00; while a complete saturation diving system with all the ancillary equipment would run in the neighborhood of $1,000,000.00. As the depth of water increases, a saturation system becomes more and more economically feasible.

Following is a typical example of the economics of saturation diving at a depth of 300 feet. The 300 ft depth is being used as an example for several reasons, among which are: (1) the majority of hyperbaric welds undertaken by Taylor Diving have been completed in the 200 to 400 ft range with less than 10 percent of all welds completed at depths greater than 400 ft; and (2) Federal regulations dictate that the cutoff depth for surface supplied diving is 300 feet.

The daily cost for surface-to-surface diving using 8 divers rather than 6 saturation divers includes surface-to-surface equipment at 27 percent of the cost of saturation equipment; labor, 75 percent; depth bonus to the divers, 171 percent; total daily cost being 87 percent of the saturation cost. But here is the difference in *actual* cost per hour of productive work: Six saturated divers can produce 22 hours bottom time per day; eight surface-to-surface divers, 8 hours bottom time per day. The cost per hour of surface-to-surface bottom time is, therefore, 240 percent higher than that for saturation diving. A surface-to-surface diver is between 65 and 70 percent as effective as the saturated diver. One of the reasons for this lack of efficiency in deep water is the encumbrance of the long umbilical and the effects of current against it. When divers are changed every hour or less, much time is lost while the succeeding diver becomes oriented. Sometimes 15 minutes or longer is required to get oriented to where the other diver left off. Taking this into consideration, the surface-to-surface diver is now costing 350 percent more per hour of productive bottom time.

As a final example, if 100 productive hours of work were to be performed in a depth of 300 feet, it would require over 15 days of surface diving, whereas the saturation divers could perform the same task in approximately 5 days. Considering that the diving support platform is usually a construction barge costing over $1,000.00 per hour, to say nothing of support vessels (such as tug boats, crew boats, and supply vessels), and the cost of support from shore, there are tremendous savings to be realized by using the saturated technique. In the deeper depths now being encountered, saturation is the only feasible method.

Along with the virtues of SAT diving, one disadvantage may be found in the cost of mobilizing the equipment, which is greater than for the simpler methods. For instance, it is *double* the cost of mobilizing a surface-to-surface system. Among the topside support personnel, more technically oriented people are required. After a crew has been in saturation for 30 days, they are generally decompressed and a new crew saturated, if the work is to continue. Whether or not the work continues, it requires approximately 24 hours of decompression for each 100 feet of depth pressure. During this time, the equipment remains on rent and the personnel can be reduced. Most of the Taylor systems are built in such a way that one crew can be undergoing decompression while another crew is in saturation continuing on with the work, thereby eliminating job shutdown due to decompression. In many instances, this type of shutdown is a major complaint on the part of the contractor when mixed gas surface diving is being advocated.

Hyperbaric Welding

Before getting into a comparison of deep diving and hyperbaric welding, a brief review is required to acquaint anyone not familiar with how the diving industry and, in particular, Taylor Diving and Salvage Company, Inc. proceeded to get involved in the art of subsea welded connections.

In the early 1960's, the founder and president of Taylor Diving saw the future need for industry to have the capability of making code-quality welded pipeline connections or structural repairs beneath the sea's surface. A research and development project was initiated to develop the equipment and procedures required to reach this goal. It soon became apparent that future requirements for the application of hyperbaric welded connections or repairs would most likely be dictated by (1) the type of weld required, (2) the reliability of mechanical connections, (3) environmental conditions, and (4) increased water depth. The types of welds required can be varied and considerable, ranging anywhere from a rather straight forward riser tie-in to that of a complicated structural repair.

The argument over the reliability of mechanical connectors (and Taylor's philosophy) is one of assurance rather than one of justification. In other words, it is possible to justify various other forms of connectors as a substitute for the hyperbaric weld, with emphasis being plainly put on the word *substitute.* Complete assurance, however, can only be provided by a welded joint that has been proven 100 percent reliable. Although there are several applications for hyperbaric welding (pipeline repairs, riser tie-ins, expansion loops, structural repairs) involving various items of specialized equipment and complexity, this paper will dwell on the pipeline connection simply because more pipeline welds have been completed and at deeper depths than any other type of welded repair. The pipe is welded in place on the seabed with the pipe ends inside a dry environment so that the joint properties will be comparable to those obtained from a conventional surface weld. This involves the use of an open-bottom enclosure or underwater welding habitat, which is placed over the pipes to be joined. This method of joining pipe together is called "habitat welding," which has come to mean "fusion welding at hyperbaric pressure by diver/welders wholly within an underwater dry environment."

The first habitat welds made by Taylor Diving and Salvage were pipeline repairs in the St. Lawrence River and in the Gulf of Mexico in 1968, made at a water depth of 60 feet.

From this first year's operation, the present hyperbaric welding process was developed and has been used since to weld all the major grades of pipeline steel and many proprietary grades. This process basically consists of:

(1) The procedure qualification

(2) The diver/welder qualification

(3) The offshore phase

The weld procedure qualification has to be carried out at the same depth or slightly deeper than the depth of the pipeline repair. This is necessary to ensure that the effects on the weld of the ambient gases and gas partial pressures are fully reflected in the finished product. The environmental conditions required to sustain life at a depth of 500 ft would be 3.0 percent oxygen, the remainder being helium, for example. This represents an oxygen partial pressure of 0.5, which compares with 0.2 at the surface in the atmosphere. This increase of partial pressure results in an increase of oxygen absorption in the weld. At 500 ft, this increase is approximately 100 percent by weight in oxygen content in the weld when compared to the equivalent surface weld.

The helium, while it is an inert gas, has a thermal conductivity six times greater than air. Therefore, to prevent very rapid cooling of the weld and the heat-affected zone, which could lead to thermal stress problems such as cracking and excessive hardness areas, the pipe temperature is maintained throughout the welding process by the use of heating blankets. The entire weld procedure qualification and weld testing are observed and approved by the customer and a representative from a certifying authority, such as Lloyds Register or Det Norske Veritas.

The properties obtained from a hyperbaric weld will meet or exceed the specification requirement expected from an equivalent surface atmospheric weld.

Before a diver/welder is permitted to weld a pipe repair, he has to qualify to that particular pipe weld procedure at the appropriate depth. This qualification is carried out either in the depth chamber or in the habitat at the job site. The qualification and the destructive testing that follows is witnessed and approved by the customer and the certifying authority. This qualification is approved for a set period and is renewed by further experience or a retest.

The diver/welders are also given instruction in the use of radiographic equipment, which eventually leads to the Assistant Radiographic Technician Level 2 qualification.

Training is also available in the use of ultrasonic equipment. Such training is set up to a particular project requirement at the time it is needed.

Offshore Phase

After the procedures are qualified, the submarine pipe alignment rig, the habitat, and other equipment are mobilized for the job, based upon the joining requirement. An adequate work platform facilitates habitat welding. It must have hotel facilities for the personnel involved and a positioning system that will enable it to maintain location in poor weather. It must have a crane or davits, or both, to handle the submarine pipe aligning rig and underwater welding habitat.

When the work platform has been located over the job site, divers go down, survey the pipe, and rough-cut it with an oxyarc torch. The cut is made by first removing the concrete weight coating from the top of the pipe, cutting a hole in the pipe, and then cutting the remainder from the inside. The cut sections of pipe are then removed and recovered. Depending on the circumstances, this may involve many hundreds of feet of pipe.

The concrete weight coating, which frequently contains reinforcing rods, is kerfed circumferentially and longitudinally with a diamond-tipped saw blade over the area where the habitat will rest. The saw is driven by an underwater hydraulic pump, which is electrically powered via umbilical from the surface. After the cuts are made, sections of the weight coating are pried off and removed by the diver. Approximately eight feet of each pipe end is bared. Davits are used to place the pipe ends in line with each other. Next, guide ropes are attached to the pipes a measured distance from the ends to ensure the habitat will land over the joint(s) to be made.

After all systems have been checked out, the submarine pipe alignment rig and underwater welding habitat are lowered to the bottom with the vessel's davits. The guide ropes, having been attached to the pipe ends a distance wider than the guide eyes on the SPAR, restrict movement of the latter, yet allow small longitudinal position corrections to be made immediately before the submarine pipe aligning rig and the underwater welding habitat are placed on the bottom.

The pipe ends are grasped and raised off the bottom with clamps of the submarine pipe aligning rig by divers operating local controls. The pipes are then coarsely aligned horizontally, each end opposite the other. The clamps are moved by hydraulic cylinders in either the vertical or horizontal direction and are powered by underwater hydraulic pumps.

The habitat is now lowered down over the pipe ends and seals are established. Plugs are inserted in the pipes and inflated. The bubble of gas in the top of the habitat is increased to below the habitat floor, displacing the sea water that was within. The environmental control system is energized and all systems are checked. Video cameras are activated to monitor the activities in the habitat.

The pipe ends are accurately aligned with the submarine pipe aligning rig clamps through the use of remote controls located within the habitat. The pipe ends are cleaned of the mastic corrosion barrier thereon, then beveled for welding with a hydraulically powered orbital milling machine. Within the habitat, a short piece of pipe called a "pup" has been stowed. This pup has been very carefully selected for circularity, examined ultrasonically, and prebeveled for welding. The dimensions between the cuts of the milling machine on the pipes are made to match the length of the pup, plus allowance for weld shrinkage. The pup is then placed between the pipe ends and held there by line-up clamps, which are used to reform the pipe ends to concentrically match those of the pup. After alignment is obtained, preheating equipment may be installed, depending upon the weld procedure. If used, the root and initial filler passes are deposited while maintaining the preheat and are made without removing the line-up clamps. The remaining fill passes and cap passes are usually applied without the line-up clamps in place. Postheating, when required, is regulated automatically without the presence of divers in the habitat.

The habitat environmental control system is used to maintain safe and comfortable conditions for the diver. The breathable gas environment facilitates fit-up of the pup since the divers need not wear masks. From the onset of welding through to completion, the wearing of masks is required.

After the welding is completed, each joint is radiographed. Three or four overlapping exposures are made circumferentially around the joint using a small, remotely-operated exposure device. The exposures are made after the diver/assistant radiographers temporarily leave the habitat and are completed before they return. The exposed film is transported to the surface for developing and interpretation.

After gamma radiography, the pipe is wrapped many times with plastic tape or cloth that has been soaked in catalyzed epoxy. The weight coating is seldom replaced, but if protection is desired, the catalyzed epoxy can be applied in water. The habitat equipment is stowed away and the end seals cut, flooding the habitat to about half its height. The pipeline is then lowered to the sea bottom and all of the clamps are released. The submarine pipe alignment rig and underwater welding habitat are raised to the surface, placed upon the work platform, and readied for the next job.

The greatest economic value of habitat welding occurs with new pipeline construction, because the integrity of a welded joint eliminates potentially high maintenance costs associated with mechanical joints. For the same reasons, habitat welding is advantageous for pipeline repairs and likely the only "universal" joining method able to be used on a large variety of pipe sizes and repair situations. For such repair contingencies, a prequalified weld procedure would significantly reduce the time a flowing pipeline would be out of service or a structural platform would be restricted. As for welder qualification, jurisdictional authorities have permitted qualification of experienced diver/welders on site to facilitate such matters.

To date, Taylor Diving and Salvage has completed approximately 200 hyperbaric welded tie-ins. In service, not one weld has ever leaked or required any subsequent maintenance. Approximately 20 percent of the welded tie-ins have been pipeline repairs. Anchor damage has accounted for 40 percent of repairs; replacement of leaking couplings and mechanical connectors by welded tie-ins for 35 percent; defective material, 15 percent; and pipeline buckles and unknown causes, 10 percent.

The maximum size of pipeline presently being welded is 48 in. in diameter. Fifty-four inch diameter pipe is within the present equipment capability, subject to minor modification. This modification is the increasing in size of certain components that have been purposely kept small to reduce helium gas consumption. The maximum depth proved in open sea tests is 1036 feet. The ultimate depth that hyperbaric welding can be used is limited by the diver's physiological limits. Although a diver's work output at extreme depths is somewhat reduced, this is not considered to be a limiting factor.

Welding at these extreme depths, therefore, will probably be done under atmospheric conditions by working inside a pressure vessel attached to the pipeline.

Most of the atmospheric welding systems so far designed and developed require the permanent installation of a pressure vessel on the pipe. While this is satisfactory for new construction where the location of the weld is previously known, this system is not suitable for the random location of welds such as required for pipeline repairs. The atmospheric welding system being fabricated and developed by Taylor Diving and Salvage has been designed for new construction, and specifically, pipeline repairs. This is achieved by using a split pressure vessel that closes over the weld area. The vessel is evacuated to atmospheric pressure for the welding process, and then after the weld completion and nondestructive testing, the pressure vessel is flooded to depth pressure, opened, and lifted clear of the pipe.

The system is being designed initially for 1,000 feet with the use of divers for carrying out preparation work such as concrete cutting and equipment setting, and will ultimately be capable of diverless operations to 2,000 feet with the use of a special submersible and remote control equipment.

Conclusion

The challenge to perform coded welds at depths greater than those presently established is one being confronted every day. Continuing research is being undertaken to overcome the various obstacles. The comment could easily be made that man has successfully spent considerable time and performed limited tasks at pressures exceeding that of the deepest laid pipeline. As encouraging as this may sound, it does not present a true perspective of the actual situation. In order to do that, it is necessary to provide certain factual information, which has been acquired through past experience. For example, there have been very few actual offshore working dives at depths greater than 1000 feet. I can cite two with which I am familiar: (1) the setting of Shell Cognac Platform and (2) the hyperbaric weld already referred to. Although both dives were undertaken in relatively the same time frame and by the same company, each required a considerable amount of predive preparation. In the case of the hyperbaric weld, it was four years from concept to actual completion. Therefore, from this example, it becomes quite obvious that deep diving is not done on the spur of the moment or on a call-out basis. Then again one might ask the definition of deep diving, and how deep is deep. A hundred years ago, sixty feet was deep. Fifty years ago, one hundred fifty feet was deep. Twenty-five years ago, three hundred feet was deep. As recent as five years ago, who, but a very few, would expect a diver making his first saturation dive offshore to be working at a depth of one thousand feet? Also, who would have thought it routine for divers to spend anywhere from 100 to 200 days a year confined under pressure? Today, this is not the unusual, but the norm. What about tomorrow? Will deep be 3000, 4000, 5000 feet? For today, 2000 feet is deep, and this is where the immediate challenge lies. This challenge is not only being met by such companies as Taylor, but also conquered.

Keeping it Going Economically: Preventive Maintenance

David Groves
Brown and Root (U.K.) Ltd.

Preface

In the early days of underwater inspection, the diving contractor was given little guidance from the platform operator regarding the objectives of the inspection task. The result was that underwater inspections, costly in both time and money, did little to assure the platform operator or governmental bodies of structural integrity.

The many conferences, seminars, and technical papers devoted to underwater inspection since 1976 have fostered the realization that the continued safety and operation of an offshore structure can be assured economically by logically planning underwater inspection and maintenance tasks, and by ensuring that details of the tasks performed are well documented and filed in an easily retrievable manner. Failure to plan ahead for inspection and maintenance could result in:

(1) Structural design details that create additional inspection and maintenance tasks or hinder their execution

(2) A critical structural member not being regularly inspected, which could result in major structural damage

(3) Inspections being inefficient and costly as two or more inspections may be carried out in the same area at different times

(4) Unplanned inspections, which disrupt platform schedules

(5) Hasty inspections, whereby inspection data is inadequately recorded and often insufficient relevant information is gathered

(6) Unreliable underwater equipment being utilized, thereby invalidating surveys and inspections made

Brown and Root's approach to underwater inspection and maintenance has undergone a logical development since 1976 and now reflects well-proven quality assurance techniques.

The aim of this paper is to outline Brown and Root's approach to underwater inspection and maintenance, which has as its ultimate goal the minimization of costs.

Introduction

Offshore structures will progressively deteriorate during their design life if no preventative measures are taken. In order to keep the oil or gas field profitable by unimpeded production and to minimize risk to operating personnel and environment, the oil operator is required to maintain the structure both above and below the water line.

With the greater emphasis on planned underwater inspection and maintenance during the past few years, it is now relatively rare for an oil platform operator to embark upon underwater inspection without clear objectives in mind. However, the planning of inspection programs is still often left to a very late stage, i.e., after installation of the structure, at which time the underwater engineer is rarely able to modify the design of the structure to either reduce inspection and maintenance requirements or facilitate the execution of the inspection and maintenance (I and M) program.

Brown and Root utilizes a three-stage approach to underwater I and M that substantially reduces I and M costs.

Stage 1: Eliminate or Facilitate Inspection and Maintenance Tasks by Improvements in Design

During the conceptual design of the offshore structure a list is prepared of all underwater items and areas that require inspection or maintenance, or both. Each area or item is examined for ease of access, frequency of inspection and maintenance, and type of underwater equipment required for accomplishing the I and M task.

The objective at this stage is to consider design changes or incorporation of special design features that will eliminate, reduce, or facilitate the execution of I and M tasks. No I and M problem is left unresolved.

Stage 2: Establish Inspection and Maintenance Philosophy and Strategy

An I and M philosophy and strategy is agreed upon between the platform operator, certifying authority, and inspection contractor. This includes the firming of manpower organization, establishing communication routes, defining acceptance criteria for defects and satisfactory completion of maintenance tasks, and establishing key milestone dates in the I and M program.

The main objective at this stage is to satisfy governmental I and M requirements for the renewal of the Certificate of Fitness.

(It is considered that the governmental requirements are general to all offshore structures and do not take into account local environmental factors or design peculiarities; therefore, the Certificate of Fitness requirements are viewed as the minimum requirements to be met.)

Stage 3: Establish a Detailed Inspection and Maintenance Program

A detailed I and M program is established and agreed upon with the platform operator and certifying authority and is designed so that by the time the Certificate of Fitness needs renewal, all governmental I and M requirements have been met. The program is also made flexible, in that I and M tasks can easily be rescheduled, so that there is adequate contingency for bad weather and equipment malfunction.

It is essential that the program is agreed upon well in advance of the starting date so that the selected diving contractor has an opportunity to review and comment on the program content and schedule.

Brown and Root (U.K.) Ltd. has utilized this three-stage approach successfully on a variety of offshore structures, including steel jacket oil platforms, submarine pipelines, semisubmersibles, and bouyant tension legged platforms.

The remainder of this paper gives details of the work undertaken in each of the three stages and generally outlines Brown and Root's philosophy of underwater inspection and maintenance.

Stage 1: Eliminate or Facilitate Inspection and Maintenance Tasks by Improvements in Design

Unfamiliarity with the problems of underwater inspection and maintenance requirements of offshore structures on the part of structural design engineers can make the tasks more onerous than need be. Therefore, during the conceptual design of the structure, it is imperative that all subsea items are examined to establish possible failure mechanisms and maintenance requirements. Consideration must also be given to the consequences of credible accidents, including vessel collisions, and damage due to environmental factors such as storms and, in susceptible locations, earthquakes.

Each failure mechanism and maintenance task is first examined with a view to its prevention or elimination by modifying structural design. If this is not possible, an inspection and/or maintenance requirement is assigned to it, and further work is carried out to establish:

(1) The importance of the item subject to failure or maintenance. The consequences of failure or lack of maintenance are examined and the importance of the item as regards safety and continued operation of the structure is assessed. The importance of the item will determine the frequency of inspection and maintenance, and even whether the item is of sufficiently low importance that it's demise can be tolerated.

(2) The available access to the item concerned. All modes of access to the item are considered including air and saturation diving, manned submersibles and unmanned tethered submersibles, and one atmosphere suits, together with various launch techniques. The selection of the most appropriate access mode and launch technique takes into account the flexibility of the diving system, depth, current, weather profile, cost, safety, power restrictions, and efficiency (i.e., diving time compared to standby time). In the case of inspection tasks, consideration is also given to the use of monitoring devices permanently situated on the structure and interrogated from the surface, i.e., use of corrosion monitoring half cells, strain gauge attachments, etc.

The underlying approach to determining the appropriate access mode is the avoidance, where possible, of divers, as there is no point in unnecessarily exposing men to potential danger. Unfortunately, the most cost-effective method to date of achieving overall oil platform inspection in waters over 150 feet deep is bell diving from a diving vessel, preferably through a moon pool. As it is not possible to eliminate the diver from inspection and maintenance tasks on oil platforms, consideration must be given during conceptual design of platforms as to whether diving will take place off-platform or from a monohull or submersible vessel. However, diving from monohull or submersible vessels can be expensive compared to diving from the platform deck, as the day rate of the diving vessel is the major part of the overall diving costs. In addition, bad weather conditions can prevent diving and add further costs, especially in the case of monohull vessels. Platform diving, with the diving spread accommodated on the lower deck and the diving bell suspended on a monorail system located under the deck, gives access to the majority of the platform structure; it is necessary, however, to use a guide weight system to lower the bell through the bracing structure. Other considerations include lowering the diving bell on a guide rail system down a jacket leg, although access would be less than that provided by the underdeck monorail system. While the proposition of platform diving may appear attractive for a single platform, it may not be economically feasible where more than one platform exists in a field.

(3) The inspection and maintenance equipment that will be necessary for the task. Failure prevention in subsea items depends upon underwater equipment to detect the impending failure in its early stages. Previous underwater trials and the experience of oil companies have shown that a large percentage of the underwater equipment currently available does not meet the manufacturers' claims, especially in the field of reliability. Often the deficiencies of the equipment are not fully realized until the inspection dive is in progress or until after the inspection program has been completed. The cost of abortive dives and uninformative inspection programs is considerable, without considering the implications for structures that are pronounced sound based solely on results supplied by unreliable equipment or the interpretation of an inexperienced diver. It is, therefore, important to select equipment that is capable of carrying out the inspections required and has a proven track record of reliability. In view of the vast range of purpose-built underwater equipment now in use or nearing final development, the responsibility placed on the inspection contractor operator to select reliable equipment and qualified personnel for surveys and inspections is considerable in terms of inspection costs.

The same comments apply to equipment selection for maintenance work, although a greater emphasis is placed upon efficiency and, hence, cost saving.

If no equipment is commercially available for the inspection or maintenance task, the issue is raised with the certifying authority in order to reach an agreement regarding a lowering of inspection or maintenance requirements until suitable equipment is available or can easily be developed for the task. Although this resolution seems a backward step, there is little point in specifying an inspection or maintenance requirement if it cannot be accomplished.

(4) Method of navigation and identification. The ease of locating and identifying items or areas on structural members underwater decreases with the decrease in visibility. Consideration must be given to navigation and identification during conceptual design to eliminate these problems during inspection and maintenance.

Vertical navigation is rarely a problem using depth gauges or the diver's pneumo. Horizontal navigation can either be accomplished by using painted lines or numbers on the structural members, or by using acoustic transponders.

Bracing nodes are probably the diver's biggest problem, as it is often difficult for a diver to determine which node he is looking at and even more difficult to determine the bracing stub identification. The problem is usually alleviated or resolved by:

(a) Permanently attaching a node and bracing stub identification prior to platform installation. The identification system must be acceptable from a stress-interaction point of view with the node, must not accelerate corrosion of the surrounding steel, and must not be obscured by marine growth. (The installation of the identification system must be monitored closely, as a mistake prior to platform installation could cause havoc later during underwater inspections.)

(b) Continuously monitoring the position of a diver by topside personnel using a scale model; the diver's location being determined by fitting on him a transducer and tracking him using transponders. (This system can be supplemented with video pictures from cameras attached to the diver's helmet.)

If method (a) is decided upon, the identification system needs to be agreed to early so that the attachment of identification tabs or painting on of numbers can be arranged prior to load-out.

In addition to identification of node members, divers also experience problems in the actual performance of nondestructive testing. Invariably they require both hands to operate equipment and are required to maintain their position against the current. It seems ridiculous that construction yards are forced to remove all fabrication aids from the jacket (such as pad eyes) when these aids may provide the means to the diver to maintain his position underwater. This is not to say that all fabrication aids should be left; clearly other considerations such as stress concentrations or the presence of sharp edges may require their removal.

Recording Format. The considerations given to inspection and maintenance tasks in Stage 1 are recorded on a standard format (see Fig. 1) so that the requirements are easily identifiable in the subsequent detailed planning of the inspection and maintenance program. In addition, a standardized format allows easy transfer to a data processing system for quick retrieval of information and sorting capabilities.

Sheet 1 of 2

<table>
<tr><td colspan="3">Inspection and Maintenance
Identification Format</td></tr>
<tr><td>Item identification:</td><td colspan="2">Category (I or M):</td></tr>
<tr><td rowspan="2">Component or Area:</td><td colspan="2">Material:</td></tr>
<tr><td colspan="2">Depth:</td></tr>
<tr><td colspan="2">Inspection/maintenance need:</td><td>Importance:</td></tr>
<tr><td colspan="3">Cause of need:</td></tr>
<tr><td colspan="3">Design impact:</td></tr>
<tr><td colspan="2">Potential I and M procedure:</td><td>Frequency:</td></tr>
</table>

Sheet 2 of 2

Inspection and Maintenance
Identification Format

Item identification:	
Mode of access:	Manning:
I and M equipment necessary:	Cost factors:
Navigation/identification requirements:	
Recommended I or M procedure:	
Comments:	

Brown & Root (U.K.) Ltd.		ENGINEERING STANDARD
		Rev:
By Checked Approved	Date	Page of

Stage 2: Establish Inspection and Maintenance Philosophy and Strategy

The inspection and maintenance philosophy and strategy should be finalized as early as possible in order to give a long lead time prior to the first year's inspection and maintenance program. Failure to do so leads to a short preliminary organization period that often results in later organization and communication problems.

Experience has shown that in the case of oil platforms, the operator's main interest after installation is getting the platform on-stream, and little time is spent considering underwater inspection and maintenance requirements until hook-up is virtually complete. Fortunately, there are certain advantages in delaying Stage 2, in that the inspection philosophy and strategy can be based upon "as-installed" drawings and, hence, reduce the problem of divers reporting "missing bracings" or identifying "unknown" items. In addition, it is likely that the platform schedules are more firm.

Inspection and Maintenance Philosophy. As the I and M requirements are already known (Stage 1), the remaining work is as follows:

(1) Review the requirements against "as-built" or "as-installed" drawings to ensure they are still valid

(2) Agree with the certifying authority and platform operator on the defect acceptance criteria for each inspection task and the acceptance criteria for the satisfactory completion of each maintenance task

The major problem experienced in early inspection and maintenance work was the lack of definition regarding acceptance criteria, i.e., the diver or topside overseer was often unsure whether a defect was significant. In agreeing upon acceptance criteria, a realistic approach must be taken; for example, it is pointless stating that all weld defects greater than 1 in. in length should be reported if the nondestructive testing equipment is incapable of reliably detecting defects less than 6 in. long.

All acceptance criteria must be carefully determined and checked to ensure that they do not infringe upon governmental legislation and thereby create problems in the renewal of the Certificate of Fitness. All infringements must be resolved by compromise before field problems are met; otherwise, there could be substantial and costly delays in the I and M program.

Inspection and Maintenance Strategy. The effort behind the I and M program must be organized; otherwise, decision making is made almost impossible due to ill-defined responsibilities of personnel and the copious paper work usually generated by I and M programs.

The operator's organization for managing the program needs to be defined so that lines of communication between the operator, the inspection team, and the diving contractor are made clear and will operate under pressure on a daily basis. Job descriptions need to be written so that personnel know to whom they report, for whom they are responsible, and the details of the function they are to perform.

A competent engineer, probably a corrosion or structural engineer, should be assigned to each inspection or maintenance task. His prime function would be to supervise and monitor proceedings and ensure proper recording of results. He should be qualified to recognize significant structural defects or deterioration on the spot. He will ultimately watch the video output on the surface and direct divers or unmanned tethered submersibles to obtain the information required by the survey or inspection program and request additional information in the case of damage observation. This engineer would also assist the operator's data recording team to analyze the inspection results to see whether the next annual inspection needs modifying in the light of current results; or he may require more detailed examination of items in future yearly inspections if he suspects a problem is developing, or less inspection if he feels the inspections are too frequent.

A procedural system must be developed for data collection and storage; otherwise, the operator's inspection and maintenance groups will find it impossible to monitor progress, assess performance, and interpret inspection data. The advantage of developing the procedural system early is that the inspection program can be modified quickly to concentrate effort on areas that are proving more susceptible to deterioration. The worst mistake that can occur in terms of data acquisition is the production of reams of data forms and miles of video tape, not collated in a form that can be interpreted. (It is considered essential to have video tapes with a date/time reference to pinpoint the place on the tape where certain information will be available.) Consideration must be given by the operator to establish a system (data processing or card system) whereby information is readily retrievable.

The various schedules for accomplishing the I and M program need to be examined from a cost point of view as follows:

(1) The cost can be substantially reduced by limiting the types of operation; e.g., it could be uneconomical to utilize air and saturation diving techniques in conjunction with manned submersibles if the cost of two support vessels must be met.

(2) The platform schedules should be examined and a tentative I and M program schedule made out to avoid delays in I and M work due to interference with other platform work.

(3) Advantage should be taken of "weather windows," primarily for inspection and maintenance tasks in the splash zone region, and secondly for the majority of the annual I and M work.

Stage 3: Establish a Detailed Inspection and Maintenance Program

Once the inspection and maintenance philosophy and strategy have been agreed upon, work can continue to establish a detailed I and M program. A separate instruction sheet for the diving contractor is made for each I and M task identified in Stage 1; the sheet lists:

(1) Purpose of inspection

(2) Parts and location of the structure to be examined with applicable drawing numbers

(3) Preinspection cleaning and NDT equipment to be used

(4) Method of access (note that this must be agreed upon previously with the diving contractor, as he is the specialist.)

(5) Description of inspection and inspection procedure

The instruction sheet also gives practical directions to the diving contractor and, hence, ensures that the overall inspection philosophy is met.

Log sheets are attached to the instruction sheets to enable the topside engineers to record directly the inspection results. The reasons why topside personnel are required to record the data directly are as follows:

(1) It is extremely difficult for divers to record information with wax crayon and slate while working with their hands.

(2) Understanding verbal commentary from a saturation diver with an unscrambler may be difficult even for trained diving personnel. Therefore, if the topside engineer hears the commentary on the spot, he can request clarification or further information if required.

(In view of the problems, the best route for information to reach topside is either via video cameras and surface readout underwater equipment, or by the use of manned submersibles with specialist engineers on board. Colored still photographs are, however, still the best method for recording structural damage for permanent records.)

The log sheets completed by topside engineers will contain all information relating to the I and M identification number, the diving contractor and divers involved, diving time log, item being inspected or maintained, video tape number, observations, and a special report in the case of observed damage. The log sheet may also indicate areas likely to warrant further investigation. The sheets not only serve as a means of recording current inspection findings, but also form a permanent record which, with other reports giving the platform inspection history, counts towards the recertification.

Each instruction sheet and the time allotted for the execution of the I and M tasks detailed are discussed with the proposed diving contractor, as it is his responsibility to perform the I and M tasks and meet the schedule milestones.

Usually the I and M program is preceded by a "base-line" survey that commences with location and removal of debris lodged within the structure and proceeds to check the structure against the "as-built" drawings in order to account for all structural members and detect obvious installation damage. The removal of debris can be a particularly hazardous task for divers due to either falling debris or the possibility of entanglement with wire hawsers, etc. Remote control vehicles should be used to locate debris so that divers can then proceed to remove it from the top, working downwards. It is unfortunate that the high loss rate of R.C.V's inside platform structures must be weighed against diver safety. Only when the "base-line" survey is complete and the danger to divers from debris is removed should subsequent inspection and maintenance programs be carried out.

The use of scale models when discussing I and M tasks with the diving contractor is recommended. In addition, divers and topside overseeing personnel can use the model to quickly familiarize themselves with the structure. The cost of the model is often saved by reduced diving time.

Underwater Repair. Inspection and maintenance programs sooner or later lead to underwater repairs. It is advisable therefore that the operator includes in his I and M organization a small team of structural engineers and a fracture mechanics consultant who are available at short notice and are familiar with the structure.

Immediate major repairs are impractical offshore as the time lag between detection and repairs must be as long as is required to accomplish the following:

(1) Obtain a full report of the damage extent

(2) Mobilize repair equipment and personnel

(3) Plan the repair operation around platform and I and M schedules

(4) Carry out computer runs to determine whether damaged areas can be removed entirely, or additional temporary bracing is required, etc.

This delay means that the first priority will be a computer run to determine the safety of the structure; therefore, the computer program and personnel must be in readiness not only throughout the structural survey, but also throughout the operating life of the platform. Failure to have this facility readily available will delay decisions that could endanger the platform and put operating personnel at risk.

* * * *

No attempt has been made in this paper to state actual costs of underwater inspection and maintenance work, or to relate Brown and Root's experience of particular diving systems and underwater equipment. The reason for this is that there are so many variables that may affect the underwater operation, including geographical location, weather profile, marine conditions, structural design, and the everchanging market for diving and support vessels. However, detailed planning of underwater inspection and maintenance tasks commencing at the conceptual design stage of the offshore structure is a logical and cost-saving approach that can always be utilized to take advantage of the latest developments in underwater technology.

Wet Welding Equipment

Herman B. Smith and Robert G. Hirsch
Norfolk Naval Shipyard

This paper deals with the effective development of methods, procedures, and satisfactory production performance in underwater wet welding in relation to currently available underwater wet welding equipment.

In late 1978, we in the Welding Engineering Division of the Norfolk Naval Shipyard were given the task of developing and qualifying an underwater wet welding procedure and then training a number of divers to qualify to it. This was necessary to accomplish some urgently needed repairs to the outer sill of our largest drydock. At the time, there were no commercially available companies who could meet our schedule requirements.

Our investigation revealed that there was very little concrete, meaningful data or information available in the technical area nor any readily available information on equipment for underwater welding. One obvious lack was a comprehensive welding standard to work to, such as the proposed AWS underwater welding specification now nearing completion.

We decided to work as much as possible within the framework of Mil-Std 248, "Welding and Brazing Procedure and Performance Qualification," for qualifying procedures and diver/welders, and fabrication document Mil-Std 278, "Fabrication, Welding and Inspection for Machinery, Piping and Pressure Vessels" of the U.S. Navy. We also used NAVSEA 0900-LP-000-1000, "Fabrication, Welding and Inspection of Ship Hulls," as a reference document. The requirements of these documents are stringent, and coupled with stipulations from NAVFAC via Mr. T.J. Dawson, we proceeded.

We built a small enclosed tank (Fig. 1) equipped with watertight door, lexan viewing window, and sleeves/gloves cut from old diving suits. The tank is designed for rapid flooding and emptying (two minutes to fill, two minutes to drain). The appropriate welding lead, ground, waterproof light, weld cleaning equipment, and plate rack were installed. We used a 400 amp Westinghouse DC rectifier as our power source, and after many problems in welding, added an arc-drive unit to the set-up, which was very beneficial in reducing arc outages.

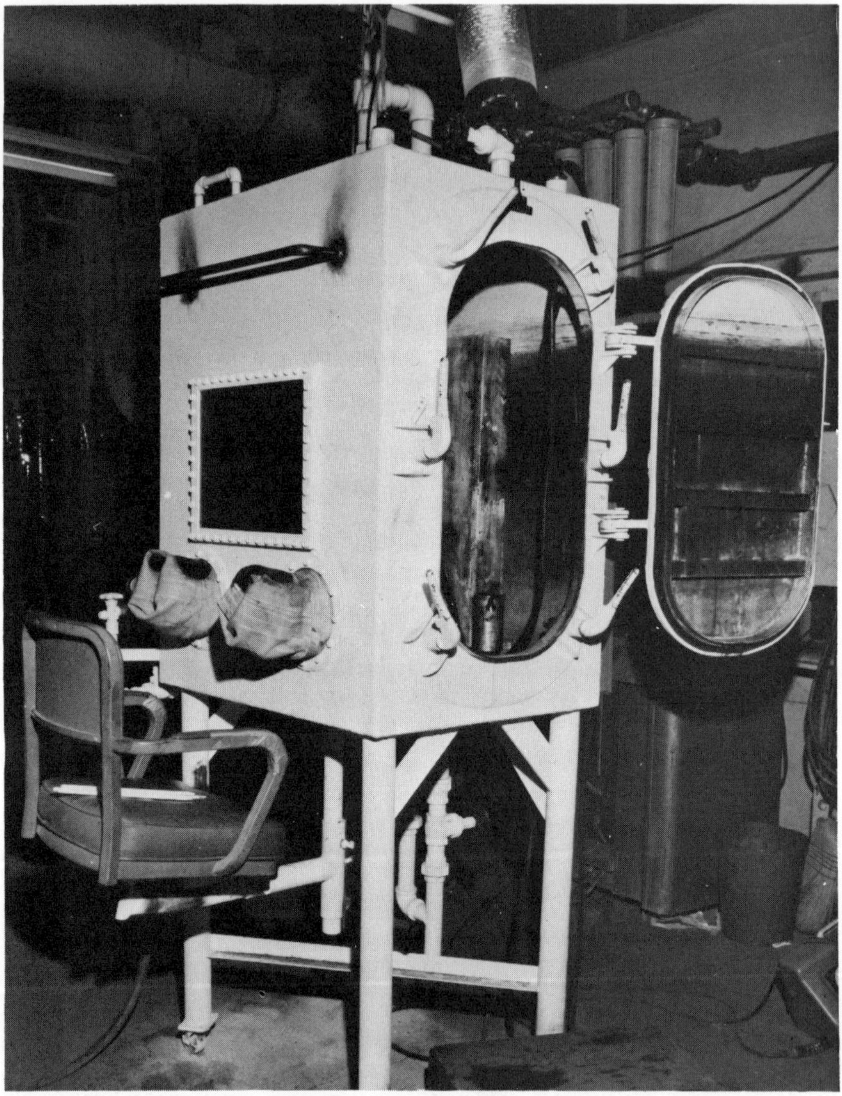

Fig. 1 – Small enclosed tank (or 'glove-box')

Using one of our lab welders and the glove box (Figs. 2, 3, and 4), we qualified a procedure for HTS and mild steel plate in the flat, vertical, and horizontal positions, obtaining results meeting terrestrial weld requirements by both nondestructive and destructive tests. Using that procedure, we trained and qualified 14 divers as wet welders, using tests in both the glove box and the river. The river work extended our qualification to 60 feet in depth, which is more than sufficient for our shore facility and ship afloat repair work.

These men have successfully completed the dry dock sill job, which has led to other work including replacing segmented waster sleeves, Seachest splitter bars, and shafting fairwaters, all waterborne at a great savings over the costs of drydocking a vessel for this work.

As to equipment, arc-drive units (Fig. 5) equal to those built by Westinghouse have proven very beneficial in maintaining a close underwater wet welding arc. That company fabricated several units for us. We also found it economical and expedient to utilize a "stinger" or electrode holder equal to the "Arc-Air Sea-Stinger." Also very useful when we worked into material prep and removal is the "arc-water" torch, proprietary to the Arc-Air Co. The divers have found this very effective.

In the area of electrodes, we regret to state that certain proprietary wet welding electrodes were found deficient in areas such as cup depth, restarting, and arc stability. We, as many others in this field, have evolved our own electrode and coating combination for best results.

Since first building the glove box, we have modified it to use clamping rings on jetline dry box gloves with accordian sleeves. This simplifies greatly the sleeve/glove replacement problem.

We also found that using soft foam rubber on the welding shields prevented scratching the lexan viewing windows.

We now accomplish overhead welding and arc-water cutting in the glove box (Fig. 6).

Use of the glove box makes possible a disciplined introduction to wet welding with guidance right at hand for diver/welder improvement, with no danger from climbing into and out of water, or handling and equipment problems. It also saves a great amount of time since immediate instructor and/or technical evaluation of progress is possible.

In summary, we have found the glove box approach to the training of diver/welders a very effective method. It is valuable both for controlled initial training and later instruction on innovative methods as they come into use. We have been very stringent in our glove box weld quality requirements, which has occasioned some biting whispers from personnel being trained; but, in every case, this discipline has paid off in the river. These same diver/welders return to thank us for our demanding approach because as they have stated, "It makes the river jobs easy."

Fig. 2 – 'Glove-box' filling, preparatory to welding vertical butt plate

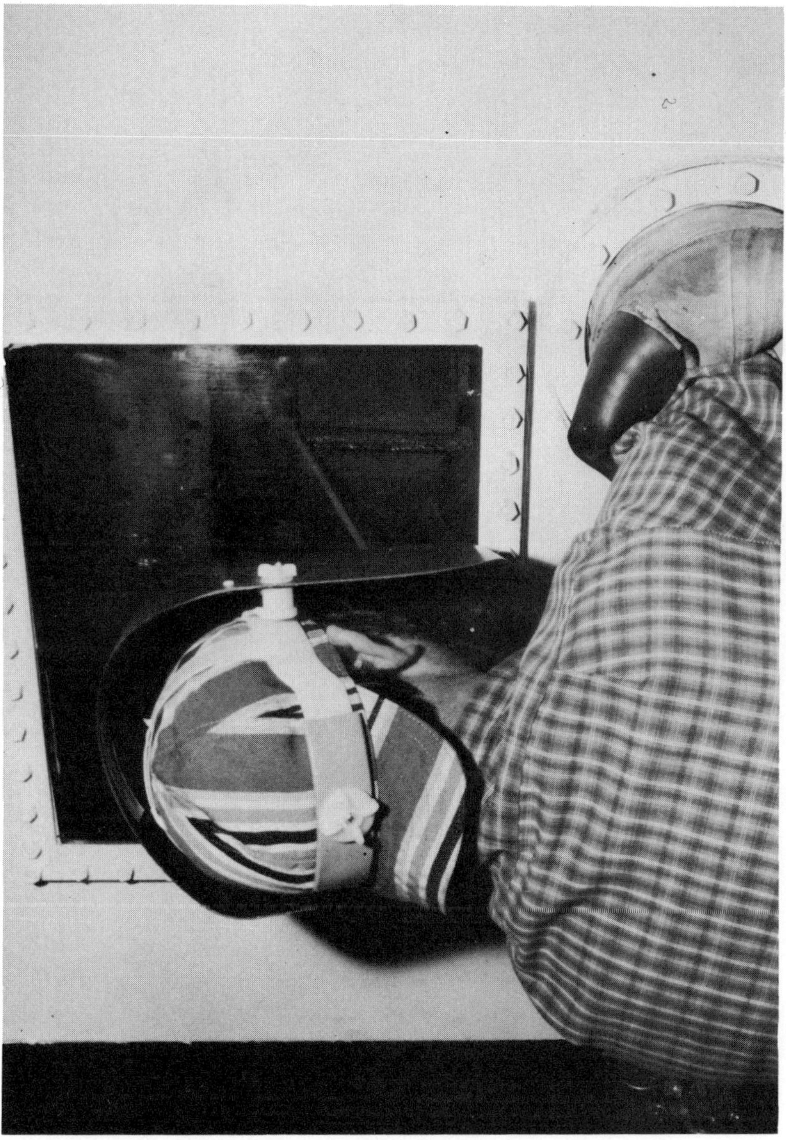

Fig. 3 – Welder ready for 'switch-on' to commence welding

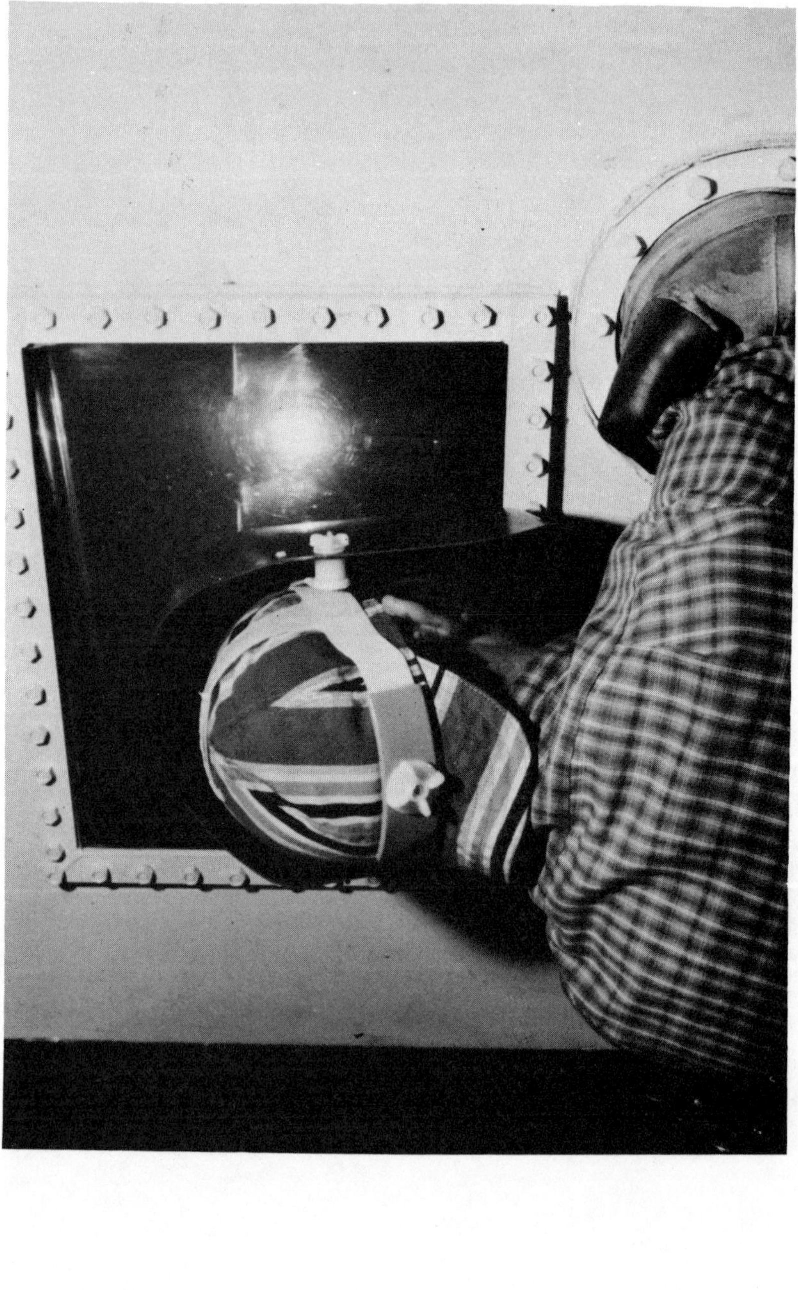

Fig. 4 — Arc initiated (note clarity of view for instructor or observer)

Fig. 5 – Arc-drive unit atop rectifier power source with volt/ampere meter unit hooked into circuit

Fig. 6 – Showing modified 'glove-box' with positive action piston pump added together with proper gauges and by-pass to permit varying output pressure as needed for 'arc-water' cutting unit.

We have at the shipyard a large HY-80 tank (Figs. 7 and 8) with heat-treated glass viewing windows, very good pumping, and associated equipment that we plan to utilize for further training and improvements. One advantage will be the opportunity to do full-scale mock-ups of certain critical type jobs and then evaluate them through both nondestructive and destructive tests.

All of the out-of-the-river glove box and tank training is done in full compliance with U.S. Navy diving safety regulations.

Fig. 7 — Large HY-80 tank with thick heat-treated glass windows

Fig. 8 – Tank with cover, associated pumping, filtering, and other equipment

Submerged One Atmosphere Welding System

Edward M. Briggs and Robert E. Adler
Southwest Research Institute

Introduction

A one atmosphere welding system is proposed as an alternate to hyperbaric welding or mechanical coupling of pipeline in intermediate or deep water.

The development of a one atmosphere welding system for deep submerged operations is feasible with the use of current technology. The system would provide a dry, one atmosphere environment, independent of depth, to allow conventional welding equipment and personnel to perform the work. The quality of the welds would be better than those done in underwater, wet, or hyperbaric conditions. This paper reviews the state of the art in the development of submerged one atmosphere systems. It provides a detailed description of one such system with emphasis on special requirements, advantages, and disadvantages and provides a basis for system tradeoffs.

Overview of Single Atmosphere Welding

Problems associated with a single atmosphere welding system are not caused by current and projected welding practices. The method of pipe preparation, pre- and postweld heat treatment, and the actual weld itself are well enough understood for other uses not to pose new or unique problems, other than the fact that they must be accomplished on a confined area and that the welders face an atmosphere that will be continuously contaminated by the products of the welding process itself.

One atmosphere welding, of course, provides a much more favorable situation for welding the quenched and tempered steels of 65 ksi to more than 100 ksi required for deep water pipelines. Therefore, it behooves the designer to design a system that is compact, meets the appropriate codes, and can allow the work to be accomplished in a confined space. The amount of automation desired is dependent upon the size of the pressure vessel in which the operation is being performed. While it is obviously one of the requirements to be able to perform the welding service as fast as possible, it should be kept in mind by the designer that a major portion of time and expense result from preparation of the pipe, and that the actual welding process part of the operation is very small. Reductions in the welding time save very little time or expense on the overall process. Some attention should be paid to the inspectability of the weld and to the equipment necessary to inspect and recode the weld. There are a number of technical and operational problems in the single atmosphere welding that control the overall costs and economic success of a single atmosphere welding system. The cost of the welding system is nominal. In fact, operational and technical problems are overriding in their importance. Basically, the intent of this paper is to describe those areas to which the designer must pay most attention in order to have a successful and economically viable single atmosphere weld system. In order to go over these systems, it is necessary to look at the total operation.

First, of course, the failed area or the area to be welded must be located. Once it is located, certain preparations must be made. Some knowledge of the topography and geotechnic properties of the soil must be made. Once this is accomplished, the pipes to be welded must be moved into position so that the pipe welding device can be lowered upon it. Depending upon the user's interest, they must be lowered into position very accurately, or they can be left askew and slightly out of line by several feet. Most of this pipe movement can be done using lines from the support barge. Since it takes a great deal of horsepower to lift and maneuver the pipes, the use of the barge lines will always be advisable. The placing of transducers and the possible pre-preparation of the pipe, that is, cutting off jagged edges and preparing the sea floor, should be performed before a pipe repair system is lowered. Currently, these tasks are accomplished primarily by divers. They could be accomplished by manned or unmanned, tethered or untethered submersibles or work systems. However, if the designer should incorporate a device for pre-preparation of the pipe and terrain into the pipe repair total system, a savings, not only in money but certainly in time, can be realized.

Once the pipe repair chamber and auxiliary equipment are lowered onto the system, a number of new problems will face the designer. However, before lowering the chamber, some attention should be paid to the frame or frames that are necessary for aligning the pipe sufficiently so that they can be welded without a miter. Certain kinds of pipe repair will require a very heavy strong frame in order to force the pipe by overbending into the weld chamber. Here the designer is faced with two options. One is to put the frame out some distance so that the pipe is flexible enough that it does not require a powerful manipulator system in order to align the pipe. The other is to perform this operation using an integral frame, which by means of a hydraulic system will force the pipe into position. There are strong arguments for both systems. However, it should be noted that some operations will be eliminated by going with one system or the other. After the pipe weld system has been placed on the pipe and we are able to maneuver it into position, we now face some designer choices that certainly affect costs, reliability, weight, etc., of the overall system.

One of these operational choices is the method of pipe cleaning. To clean the pipe out in open water to make an O-ring type seal precludes even the most minor blemish be made on the pipe. This is a very difficult operation requiring automation and a "smart" computer controlled system. Another system to be considered is the pigging operation. Pigs must be put in the pipes in order that one atmosphere may be formed in the pressure vessel. Since pig plugs can be very heavy and must be inflated and deflated, pumped out, and inserted into the pipe with a complex mechanism, careful design processes can save considerable time and expense. The third important operational feature is the seal of the chamber to the chamber and most importantly around the pipe. We face several methods of doing this. Currently, in the literature we can find a welded seal that means potential damage to the pipe and leaves an expensive structure on the bottom with the unknown metallurgical problems. While the weld system is simple, it presents several very difficult problems if one considers operating one of these systems for a great period of time. The second system is a conventional sealing system moving some sort of captured elastomeric seal. These systems, of course, have the advantages of years of experience. However, with the very complex seal required, that is, the circumferential and longitudinal seals meeting, and with the difficulty of cleaning the large diameter pipe sufficiently until a reliable seal can be made, some difficult problems are presented. The advantage, of course, is that no residual material is left after the seal is made. To ensure against blowout, back-up seals can be placed by welders after chamber entry. Coating the pipe can present some problems since it cannot be recoated until the seal is broken. The third method is to seal by injecting some sort of material that has good bonding qualities and will set up with sufficient strength to stand the external pressure. The advantages of this seal would be more tolerance in alignment and more tolerance in cleaning the pipe. Problems exist in the fundamental nature of the R and D necessary to prove the third method will work, since experience with this kind of system in the oceans is not well defined. This sealing system will require more extensive procedures than either of the other two.

The selection of the method of pipe cleaning, pipe pigging, and welding chamber seal will have a direct impact on the cost, weight, and reliability of the system. The other major problem the designer must contend with is system safety; that is, emergency procedures must be identified at the design stage to provide both redundant and fail-safe systems to protect the personnel. The final severe problem is the pig. In any case, a pig must be inserted in both ends of the pipe in order to seal the pipe. If the pig is a hydraulically expanded rubber bladder with grippers, the designer must be absolutely sure that no leaks or failures can occur without sufficient advance notice. Pig movement must be detected and advance warning provided. There are currently no codes or regulations covering the use of an inflatable pig in man-made chambers. Therefore, it is necessary for the designer to prove to the code and regulatory agencies that this system is as safe as a normal hatch or window or other penetration normally found in a man-rated chamber. Of course, a pig that is noninflatable with a strength member between the two sealing ends and with both mechanical and hydraulic seals would probably present an easier time with the code and regulators; however, it presents severe problems with installation since it must be carried with and installed as the pup is installed in the chamber. Minor problems the designer must address are control of noxious gases and particulate matter from the welding processes and providing large amounts of power to the system. A one atmosphere mating and transfer system, of course, is necessary on all these systems. This problem has been studied and a variety of techniques have been used. The problem does not appear to be as severe as some of the other problems because of its wide exposure.

There is a depth limitation to the system that is really controlled by cost and technology. The deeper the system is deployed, fabrication and metallurgy problems begin to dominate the design. Operationally, the placement of the system and the extension of the cables and electrical umbilicals begin to become difficult. Using current technology predictions, it would seem that a practical limit is 5000 feet. One must also consider the capable range of the system to work at shallower depths and still maintain a cost competiveness with hyperbaric welding systems at 600 feet. Single atmosphere welding systems do have limitations, but at the current time they are feasible to accomplish work to 5000 feet.

Description of a Single Atmosphere Welding System

The following system was designed to the specific requirements of Taylor Diving and Salvage Company of Belle Chasse, LA. The system is about 85 percent complete, and a detailed description is given to help the reader appreciate the problem outlined in the previous pages. An artist's conception of the SAWS is illustrated in Fig. 1. Its purpose is to permit welding on submarine pipeline in a dry, one atmosphere controlled environment. It consists of a submarine pipeline alignment frame (SPAF), several one atmosphere chambers, pipe manipulator system, and necessary hydraulic pneumatic, electrical, electronics, life support, and environmental control subsystems to support the welding operation and the life of repair personnel. The system is designed to operate at depths to 2000 feet. The system will meet the manned submersible requirements of the American Bureau of Shipping.

The total weight of this system is 400 short tons in air and 100 short tons in water. It was designed to be supported by special derrick/pipelay barges available to Taylor Diving and Salvage Company. To maintain buoyancy control and to provide a footprint pressure of less than 85 psf, the tubular SPAF will be pressurized to dive depth pressure.

The barge will be able to place the SPAF within ±5 feet of the repair location of the pipe to be welded.

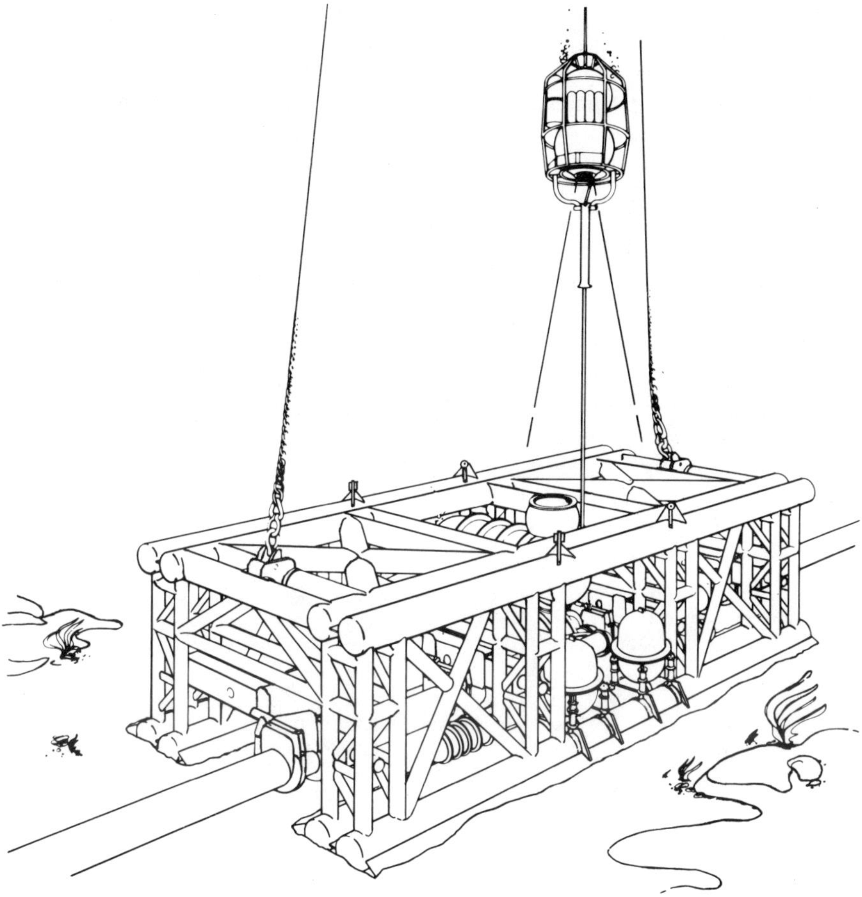

Fig. 1 – One atmosphere welding system

The SPAF consists of four transverse frames forming three longitudinal bays. Within each frame is a hydraulically operated pipe claw and traversing mechanism (screw jacks). Semi-rigidly attached within the SPAR are the man-rated, single atmosphere chambers and interlocks shown schematically in Fig. 2. After emplacement of the SPAF, personnel are transferred to the auxiliary chamber (AC) by a tethered one atmosphere personnel transfer chamber (PTC), and entrance to the AC is accomplished by means of an interlock scheme. The buoyant PTC descends to the SPAF via a haul-down cable, which is attached to the SPAF, and the cable is collected on a winch mounted on the exterior of the PTC. The PTC is then latched to the "tea cup," and a seal is accomplished so that the tea cup can be dewatered. When water is cleared from the tea cup and pressures in the PTC, tea cup, and AC are equalized, personnel can open hatches and enter the AC to begin work.

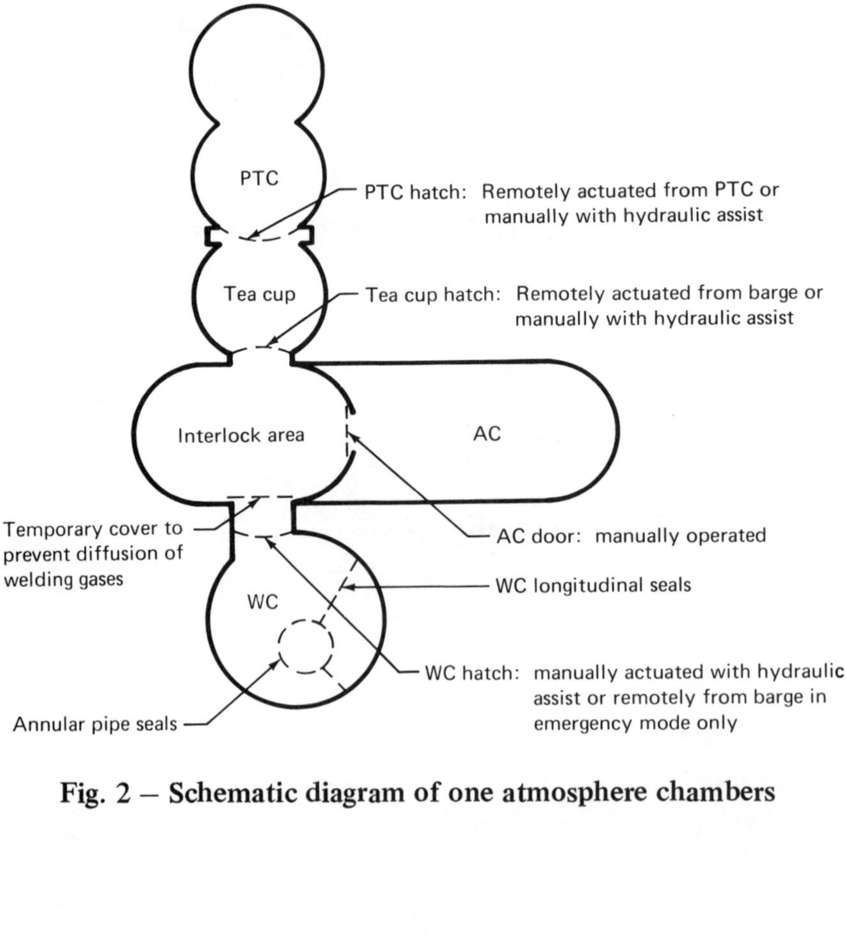

Fig. 2 – Schematic diagram of one atmosphere chambers

Before the SPAF leaves the barge, the welding chamber (WC) will be opened so that its two sections may straddle the pipes when the SPAF is placed on the sea floor. After SPAF emplacement, an operator in the AC works the pipe claws to "nibble" and finally "grip" the pipes, then moves each pipe sequentially to the "pipe cleaning"/"plug insertion" stations. At these stations, the inside of each pipe is cleaned and pipe plugs (PIG) are inserted and "set" or energized. Following PIG insertion, each pipe is moved into the fixed half of the welding chamber. The extremely critical pipe alignment is accomplished by means of a coarse and fine stage pipe-position monitoring scheme.

After both pipes have been aligned in the fixed part of the WC annular seals, the WC door is closed and secured. The pipe is then gripped with a series of wedges to prevent axial motion of the pipe, after which the pipe alignment is checked and readjusted if necessary. The annular pipe seals are then actuated. Next the longitudinal WC door seal is "set" by creating a positive hydraulic pressure differential on the lips of the longitudinal seal. After the WC is dewatered, personnel may enter the WC to prepare the pipes for accepting a previously prepared pipe splice section, which is stored in the AC interlock area. Using a small hoist, the pup is removed from its stored position in the interlock area and hoisted through the WC hatch into the WC. Another small hoist within the WC is used to orient and position the pup for welding between the two pipe ends. In the event that the pup does not fit adequately between the pipe ends for welding, a second pup will be prepared to the required dimension on the barge and transported to the SPAF via the PTC.

Two heating elements are attached to the pipes to bring the materials to the correct preheat temperature prior to the welding operation. Before welding starts, pressures in the AC and WC are adjusted so that the WC pressure is approximately 0.1 psi below that in the AC, and a temporary shield is placed over the WC hatch to restrict the flow of contaminated air from the WC into other chambers during the welding operation. Two welders may work simultaneously and will wear breathing masks supplying filtered surface air. After the welding operation starts, any other personnel entering the WC will also wear breathing masks.

Following the completion of the welding operation, the weld joints will be nondestructively inspected by x-ray or ultrasonic methods in accordance with customer specifications. Equipment for these inspections will be transferred from the barge to the SPAF via the PTC. After acceptance of the welds, the pipe surfaces within the WC will be coated for corrosion protection. Following that, personnel will exit the WC with their equipment and close the WC hatch. The pipe plugs (PIG) will be de-energized, and afterwards the WC will be flooded. The AC operator will then open the WC door, lay the repaired pipe to the ocean floor, and shut down systems prior to returning to the barge in the PTC.

SAWS Major Functional Subsystems

Electrical. The electrical power system will have three phase loads. Voltages will be 2400/4160 V, 277/480 V, 120/208 V ac and 24 V dc. The system will receive power through a power umbilical and transform it by means of two transformers located on the SPAF. An emergency and auxiliary battery power system for lighting, communications, and life support functions can supply power for a 24-hour period.

Life Support. Life support will be provided to the personnel transfer chamber, the auxiliary control chamber, and the welding chamber. It will consist of the following subsystems:

(1) Environmental control
(2) Mask breathing
(3) Food and drinking water supply
(4) Waste disposal
(5) Smoke removal
(6) Gas diffusion prevention

The system will have a capability of 1,200 mission hours. The environmental control system will maintain a working environment by the use of:

(1) Pressure monitoring and control
(2) Oxygen replenishment monitoring and control
(3) Carbon dioxide scrubbing monitoring and control
(4) Temperature monitoring and control
(5) Humidity monitoring and control

A smoke removal subsystem shall dispose of the smoke generated by welding by means of fume filtering units and ozone filter units.

Note: The following gas levels will be periodically measured in each man-occupied chamber. The minimum detectable concentrations are shown below.

(1) Phosgene ($OCCl_2$): 0.05 ppm
(2) Phosphine (H_3P): 0.1 ppm
(3) Hydrogen (H_2): 0.5% by volume
(4) Carbon monoxide (CO): 5 ppm
(5) Ozone (O_3): 0.1 ppm
(6) Nitrous fumes (NO_2): 1 ppm

The welding system shall provide the equipment, machines, subsystems, etc. required for welding in the WC. The welding system consists of the following subsystems:

(1) Power units: Five welding machines; four power supplies housed in WC (two for preheating, two for welding), one power supply stored on the barge. (If needed, the latter unit would be transported to AC interlock via the PTC.)

(2) Input power: 440 V, 30, 60 Hz at 50 amps

(3) Weight: 1.1 tons (approximate total for 4 units)

The weld inspection system shall document the equipment, machines, subsystems, etc. required for x-ray inspection in the WC of performed welds. This equipment will be stored on the barge and transferred to the WC via the PTC.

Welding Chamber Atmospheric Control. Welding of the pipe sections will contaminate the atmosphere of the welding chamber. A smoke removal system will be used to remove the bulk of the contaminants, but the system will not be adequate to maintain the degree of air purity required for breathing. Therefore, before welding operations are initiated, personnel in the welding chamber will be required to don surface supplied air breathing apparatus (BIBS).

The welding operations will also contaminate the atmosphere of the interlock and possibly the PTC and auxiliary chamber unless it is isolated from the interlock. Therefore, a hatch cover will have to be installed prior to welding.

The welding chamber BIBS system will be from the surface supplied air manifold. Full face masks will be used, and the air will be expired into the welding chamber's atmosphere.

An air compressor will be used to draw the air from the welding chamber and expel it into the surrounding water through a small volume isolation tank. The primary air compressor will be supplemented by a second "standby" unit.

The air compressor(s) will be controlled by a barometric switch set to load the compressor at approximately 14.7 psia and unload it at approximately 14.6 psia.

The pressure in the interlock/auxiliary chamber area will be kept at four to six inches of water above 14.7 psia at all times by connecting the reference port of the differential pressure switch controlling the air compressor to the pressure of the auxiliary chamber. The positive pressure in the interlock/auxiliary chamber should preclude diffusion of the welding gases.

A smoke removal loop containing two filtering units will be used to remove the bulk of the welding smoke and fumes from the welding chamber's atmosphere. The filter units will be cascaded with three ozone filters.

Whenever personnel pass through the manway and open the isolation cover, the positive pressure in the interlock area will tend to equalize to the pressure of the welding chamber. Therefore, a make-up air system supplied from the umbilical air will be provided to reinstate the auxiliary chamber's pressure. Also, whenever the pressures tend to equalize, the pressure differential switch will load the compressor(s). The make-up air can be derived from the on-board secondary air bank in case of temporary umbilical air line failure.

The effectiveness of the antidiffusion system (WC hatch cover and AC positive pressure) will be monitored by periodically checking the interlock area atmosphere for the presence of toxic gases with a hand-operated gas detector unit. The following gases will be monitored:

(1) Carbon monoxide

(2) Nitrogen oxide

(3) Nitrogen dioxide

(4) Phosphene

(5) Ozone

Sealing System. The most critical and unique system is the weld chamber sealing system. The welding chamber provides the space, environment, and means for welding and inspecting welds on pipes from 12 in. to 36 in. O.D. in a dry, one atmosphere underwater habitat. The chamber weighs 30,000 pounds in air and has internal volume of 520 cubic feet. It is an HY80 steel, one-inch thick chamber consisting of a cylinder with two hemispherical ends. The inside diameter is 8 feet, and it is 14 feet from end to end.

The chamber is hinged to accommodate a longitudinal neoprene rubber seal, neoprene annular seals for sealing the ends of the chamber around the pipe, and actuator assemblies to set the annular seals under hydraulic pressure.

The longitudinal seal prevents water intrusion to the chamber from between the plane surfaces of the WC door and WC shell. Initial contact of the seal is aided by applying water pressure between the opposing seal lips. After initial contact, the inner lip is self-energized by pressure differential.

The annular seal consists of a number of segments. Sealing action is performed by the use of hydraulic actuators that deform the neoprene, forcing it into solid contact between the weld chamber and the pipes. After the seals are set, the welding chamber is pumped out to one atmosphere pressure.

Future Developments

Submerged one atmosphere welding is the only way to weld underwater with a high degree of quality control. At great depths, it may be the only way to weld. The system discussed is particularly designed for pipelines and riser connections. A single atmosphere welding system is basically a manrated chamber that is placed in the proper position, sealed, and evacuated to allow access by a submersible or diving bell for entry of qualified welders and equipment. The initial investment cost of the system will, of course, depend upon the nature and complexity of the welding work to be performed. The operating costs of the system can be amortized over the life of the system and would be less expensive than direct underwater diver welding or hyperbaric welding. The cost differential between the one atmosphere system and other welding methods can qualitatively be compared in that no special equipment or personnel qualifications are required. One atmosphere systems can be adapted to the emerging automatic welding techniques, which would further reduce personnel requirements as long as high quality welds are obtained.

The system can be designed to fit current and future offshore barges and platforms and would require a minimal impact on redesign of existing offshore equipment.

The technology is available to build and test these systems today. Sealing concepts need further testing in the laboratory and at at-sea operations, but have proven feasible in special tests. Additional testing is required in proving the removal of critical welding by-product toxicants, but BIB control systems could be used.

The most significant system disadvantage is the initial investment cost, but this can be amortized over the life of the system since high quality welds can be made at any depth.

Advantages of the system are:

(1) Little need for additional training for qualified welders

(2) No requirement for the use of costly inert gases

(3) Work time on-station during and after task performance is not limited by decompression requirements

(4) System is reusable with minimum refurbishment

(5) Quality of welds at depths to 5,000 feet equivalent to normal surface welding technology

(6) Facilitates welding at depths beyond capability of current welding systems

(7) Reduce operating costs as a result of less time on-station, lower personnel skill requirements, and no requirement for special welding equipment or inert gases.

Hyperbaric Welding in the Repair of Offshore Pipelines and Structures

Gilbert Coriatt
Comex Services

Introduction

The exploration and exploitation of offshore continental shelf oil and gas fields are relatively new. While examples of offshore structures can be found from prior to World War II, it is over the last decade that the immense steel and concrete structures have appeared and have risen from greater depths to rougher seas and worse weather conditions.

Over this same period of time, hyperbaric welding has been developed and established as a method for pipeline tie-ins and repair and for structural repairs. Welding down to 160 meters is now undertaken on a routine basis in the North Sea, and trials down to 300 meters have proved the feasibility of hyperbaric welding for pipeline connections and repair in deeper areas such as the Norwegian Trench, the Gulf of Mexico, etc. More recently, the development of one atmosphere welding techniques has shown that these are capable of being used for increased depths.

Causes of Damage

The causes of damage for which weld repairs have been executed are numerous and include the following:

(1) Inadequate design

(2) Accident, such as impact of structures by vessels or loads inadvertantly lost, anchors and jet sleds on pipelines, etc.

(3) Corrosion of poorly protected structures, internal corrosion of risers and pipelines

(4) Fatigue

Regardless of which of these factors is responsible for the repair, it is essential that a complete and detailed survey be carried out to establish the location extent and nature of damage, any possible restrictions on access for divers and equipment, and local operational and marine conditions that may interfere with the repair. From this information, detailed engineering procedures and work schedules can be drawn up and valuable site time can be saved.

Even when significant damage to structure or line has been established and engineering analysis indicates repair is required, economic, safety, environmental, and even political factors will play an important role in deciding when and how the repair will be executed.

Welding Techniques

Techniques that have been used for underwater welding include wet welding, locally shielded welding, dry hyperbaric welding (welding in a completely dry chamber at the ambient pressure) and one atmosphere welding. In addition, techniques that have been considered include explosive and friction welding.

Wet welding has been used for a number of years for certain cosmetic or temporary repairs and attachment of minor fittings using rutile of iron oxide mild steel electrodes or high nickel electrodes. The electrode coating is protected by paint, varnish, laquer, or some other proprietary layer. The main drawbacks of this technique are:

(1) Reduced arc visibility, which requires a touch and drag technique (Consequently, manipulation of the arc and weld pool are practically impossible and welding in all positions difficult.)

(2) High levels of porosity and lack of fusion

(3) Rapid weld quench rates with high hydrogen contents resulting in poor weld metal properties

Hyperbaric Welding

Since 1973, COMEX has undertaken research and development in hyperbaric welding with the result that procedures for fillet welds and welds in butt joints have been fully qualified down to 300 meters. This research has been in two parts: small scale tests, and manned hyperbaric welding trials and qualifications.

The small-scale tests are used to evaluate the welding processes, parameters, deposition and arc characteristics, and weld metal properties at different pressures. These are carried out in a small instrumented pressure vessel fitted with a movable work table at the Institut de Soudure (French Welding Institute) in Paris.

The results obtained from these tests are then used in the development of manned procedure tests and trials in COMEX's hyperbaric center (Figs. 1 and 2). This consists of a 5-meter diameter sphere, or work chamber, which can be flooded to 1/3 its depth, simulating habitat conditions. Attached to the sphere are living chambers capable of accommodating eight welder/divers and a control cabin where all stages of the work and life support system can be monitored. Throughwall connections are available for functions such as welding, preheating, grinding, lighting, etc., and welding fumes are sucked away from the weld area and the gas(es) cleaned and regenerated. First, the

CO is catalytically converted to CO_2, the CO_2 scrubbed by soda-lime and the gas recirculated. Continuous gas analysis (O_2, CO, CO_2, etc.) is carried out and temperature humidity monitored. Similar gas regeneration systems are fitted to operational welding habitats where dry transfer and working are to be carried out.

Fig. 1 – Hyperbaric test at -160 meters in "hydrosphere"

Hyperbaric Welding Techniques. It was decided early in the development that welders would work in "shirt-sleeve" conditions without breathing masks. Consequently, GTA welding was abandoned since argon has a narcotic effect and cannot be effectively removed by the gas regeneration system. In addition, GTAW is relatively slow and subject to magnetic arc blow, possibly requiring demagnetization of the components to be welded. GTAW in helium gas was found very difficult to perform. Development has been on manual metal arc stick welding using basic electrodes and gas metal arc and flux cored welding. But we must say that other companies are successfully using GTAW for root and hot pass.

Fig. 2 – Hyperbaric center in Marseilles

Manual Metal Arc - SMAW. It is well established that increased pressure constricts the arc and can lead to instability. By modification of the coating, using a short arc technique, and increasing the arc voltage, stable arc welding conditions can be achieved at pressure. Evaluation of welding machines established that one with good dynamic characteristics and a device for arc restriking was most suitable.

For a basic low hydrogen electrode (AWS type E 8018), the increase in C content and the resulting effect on impact properties is illustrated in Figs. 3 and 4. This can be directly attributed to the increased solubility of C coming from the decomposition of the lime coating.

$$CaCO_3 \rightleftharpoons CaO + CO + 1/2\ O_2$$
$$CO \rightleftharpoons C\,(Fe) + O\,(Fe)$$
$$CO + 1/2\ O_2 \rightleftharpoons CO_2$$
$$H_2O \rightleftharpoons 2(H^+)\ Fe + O\,(Fe)$$

After evaluating several commercially available electrodes in this class, batches of electrodes with reduced lime coating were produced. In addition to restoring some of the impact properties, the electrode selected showed a low carbon content; it was less hygroscopic and had better handling characteristics and a readily detached slag. Results obtained from pipe welds at 150 and 300 meters using a 2.5 mm diameter rod with a multiple-pass stringer bead technique are given in Table 1. While the impact energies at 300 meters are perfectly adequate for all the major pipe welding specifications, comparison with results at 150 meters suggests that impact properties would not be achieved at depths significantly beyond 300 meters if low Ni content electrodes are to be used.

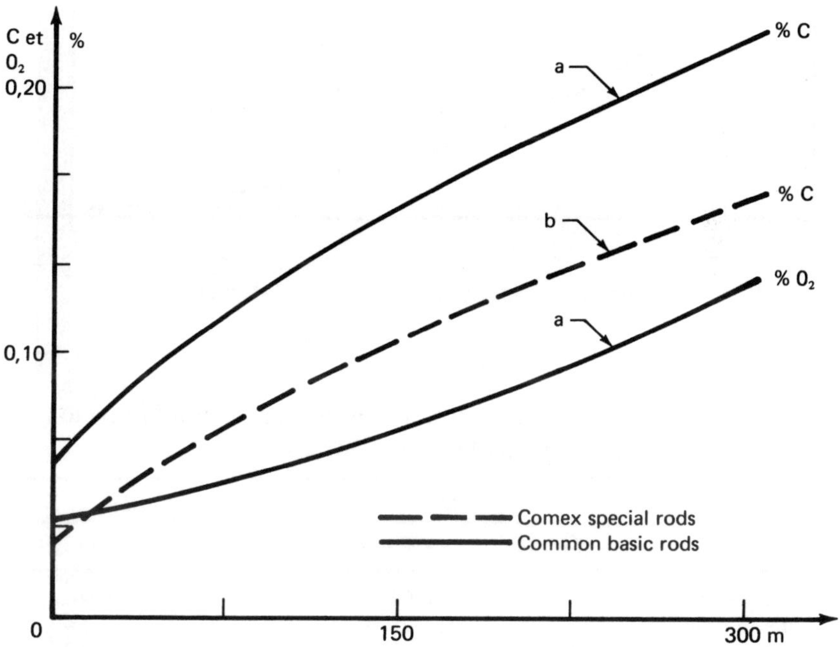

Fig. 3 – Carbon and oxygen content in weld metal versus depth

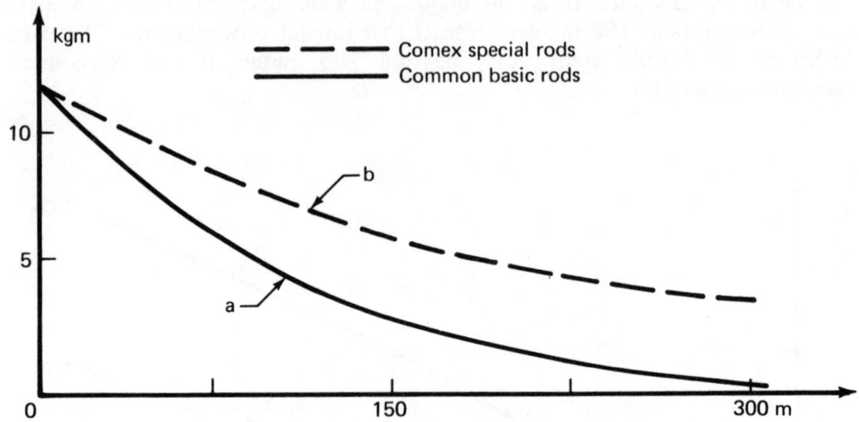

Fig. 4 – Impact properties in weld metal versus depth

Table 1
Impact results (–10° C) obtained from manual metal arc welds at 150 and 300 meters water depth

	150 M		300 M	
Position	Min (J)	Ave (J)	Min (J)	Ave (J)
Weld metal	79	91	43	46
Fusion line	68	76	50	54
HAZ	237	246	112	132

GMA Welding. Preliminary indications showed that gas metal arc welding had certain advantages over manual electrodes, particularly for repair work. First, the absence of an external coating and reduced moisture absorption means that for certain materials welding can be done without preheating; secondly, the alignment and joint tolerances are greater; finally, the process is capable of higher deposition rates with no down-time for electrode changing.

Development of the GMAW technique over the last 2 to 3 years has resulted in the approval of butt joint and fillet welding procedures down to 300 meters and a number of repairs have been carried out at depths from 5 to 50 meters.

Two of the principle factors investigated were the power sources and consumables.

Power source: Classical, flat characteristics, pulsed GMAW, and vertical characteristic sources were all evaluated, both in the small test chamber and in manned trials. By monitoring voltage and current using high speed recorders the areas of arc stability could be established with pressure (Figs. 5, 6, and 8). It was found that a welding set with vertical characteristics and independent regulation of variables was the best.

Consumables: Numerous wires (plain and flux cored) and gas mixtures have been tested. For deeper welds using a dry transfer system, a heliox shielding gas is used, allowing the welders to work without breathing masks, the gas regeneration system working in the normal way. (Even with a flux cored wire, the fumes were still less than with electrodes.) Wire combinations consisting of a 0.8 mm solid wire for the root and 1.0 to 1.2 mm flux cored wire for filling and capping passes have been developed using a vertical-up technique. The effect of pressure on weld metal analysis is shown in Fig. 7. As can be seen, there is no increase in C content or significant change in chemical analysis. Consequently, mechanical properties are only slightly affected by pressure. For example, average weld metal impact energies of 66 joules were obtained from GMA pipe welds at 300 meters (Table 2). Tensile properties, hardness, etc. were equivalent to those on the surface. Using this technique, it can be predicted that welding down to 500 meters and more should be possible in hyperbaric conditions.

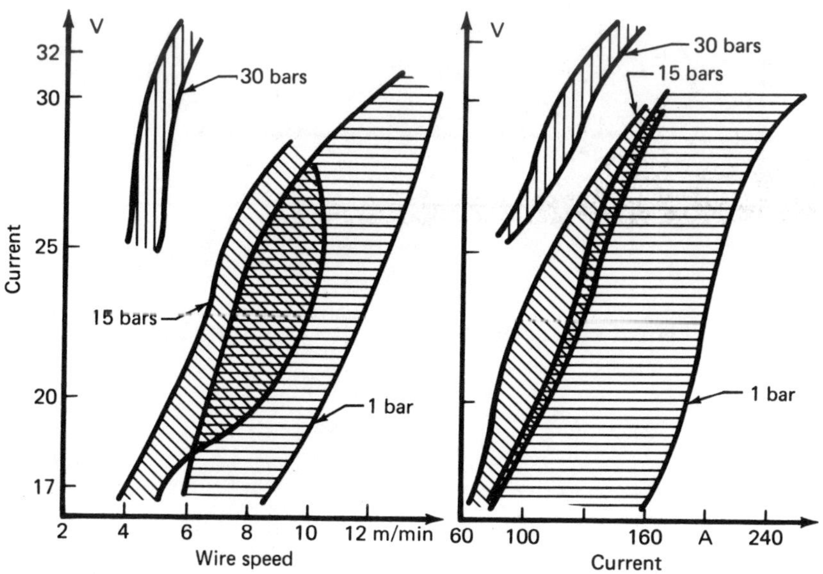

Fig. 5 – Area of arc stability in GMA welding at different depths with flat characteristics

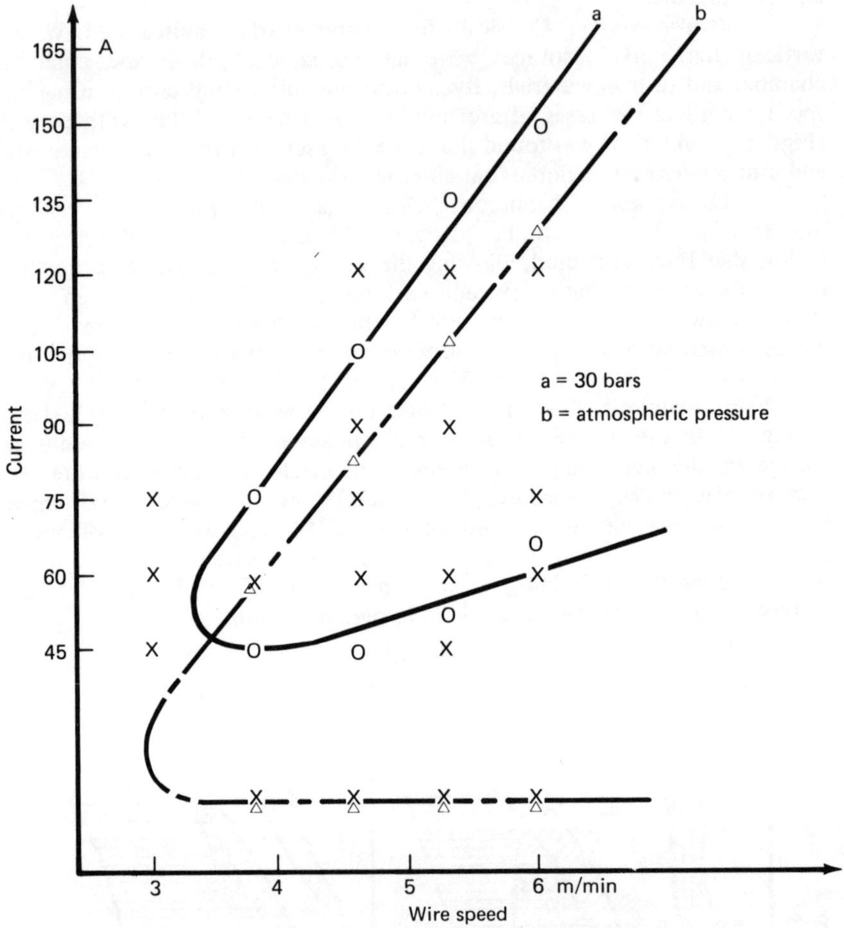

Fig. 6 – Vertical characteristics and energy controlled welding set

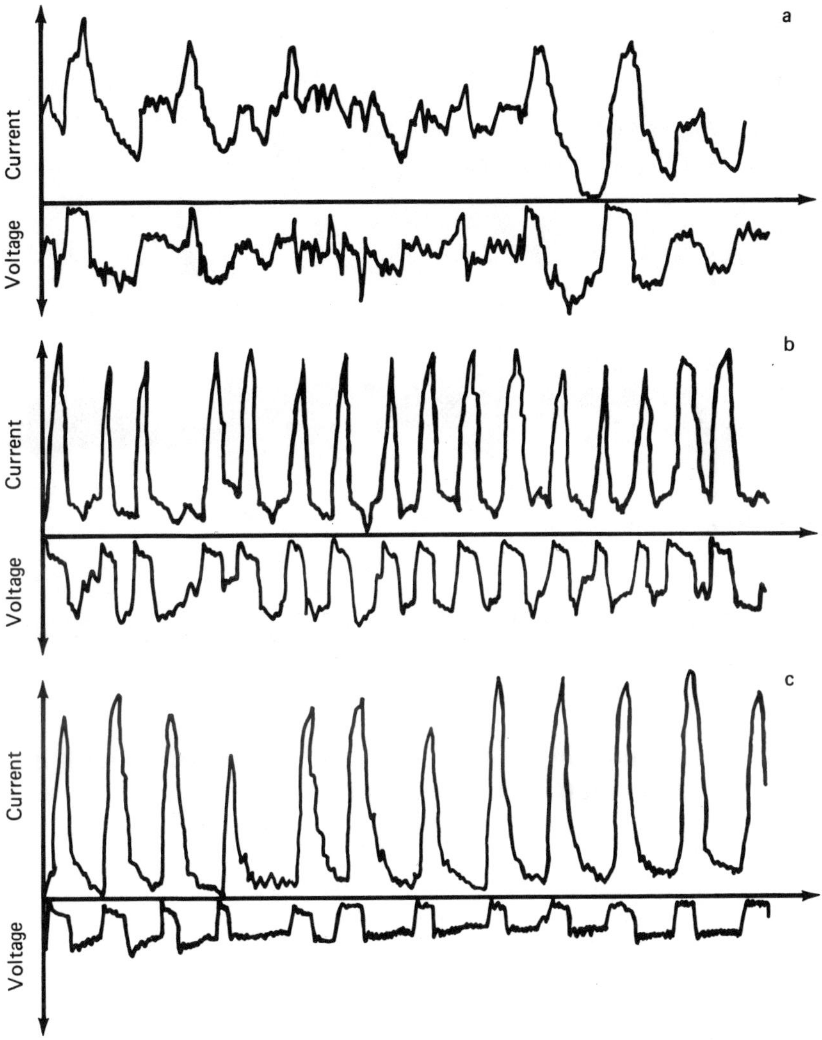

a = 30 bars
(flat characteristics)

b = 1 bar
(vertical characteristics)

c = 30 bars

Fig. 7 – High-speed records of current and voltage with flat and vertical characteristics

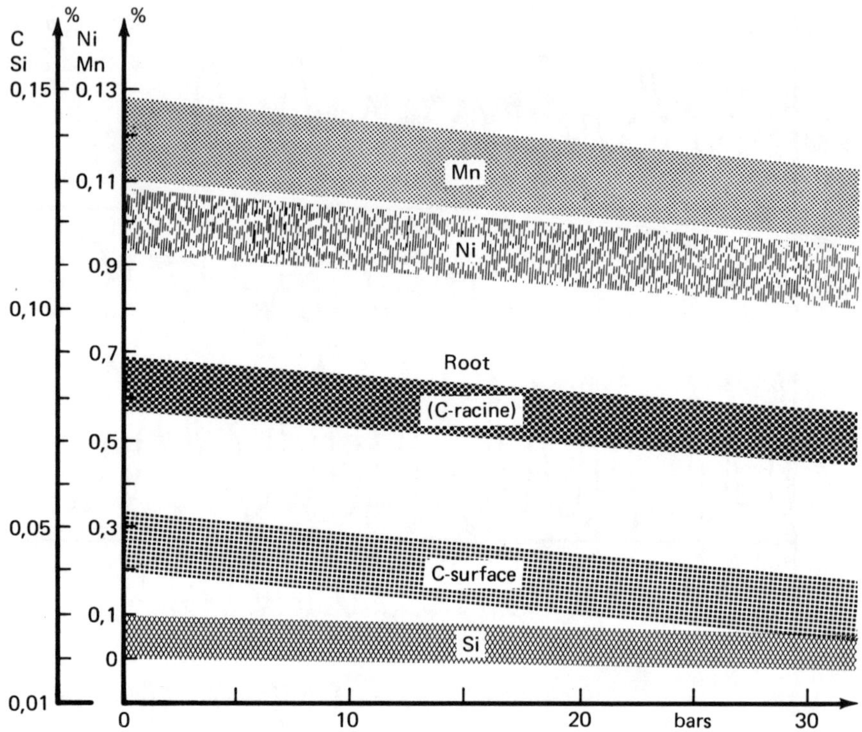

Fig. 8 – Carbon, nickel, manganesium, and silicon content in weld metal versus pressure

Table 2

Charpy-V test at minus 10° C Depth 300 m
1 mm flux cored wire
Steel: API 5LX 65 20" O.D. 13/16 wt CE 0,45

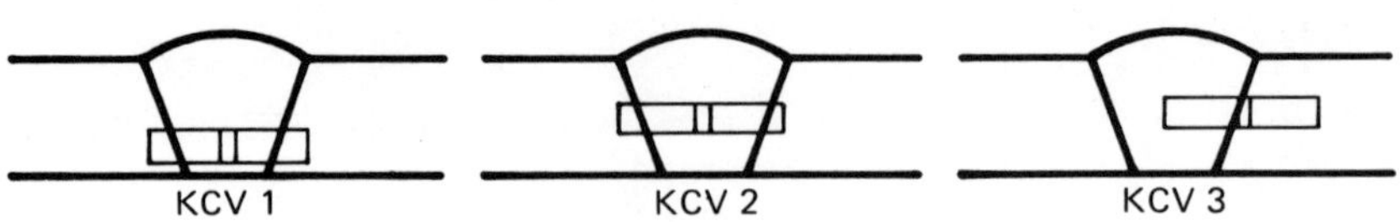

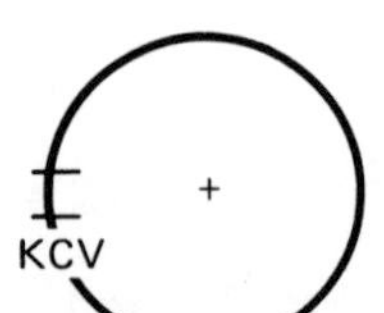

Referency	Minimum		Average	
	Kg/m	Ft/lb	Kg/m	Ft/lb
KCV1	4,8	34,2	5,3	37,76
KCV2	6,0	42,7	6,63	47,2
KCV3	6,2	44,8	10,6	75,53

Atmospheric Welding

For pipe welding at depths below which saturation welding is no longer viable, COMEX has developed a system of welding at one atmosphere called WELDAP. Development started in 1975 and in February 1978 successful pipe welding trials were concluded on a 20 inch diameter pipe at 300 meters in a Norwegian fjord. In simple terms the system is based on the use of a consumable pressure vessel or work chamber onto which is locked a second, recoverable sphere containing all the ancilliary equipment for welding, cutting, preheating, etc. Transfer of personnel is by piloted submersible, which locks on to the second support sphere.

Sealing plug and pup piece installations are performed by means of an ingenious patented system. The welding chamber is flooded and the pup piece is automatically set between the two ends of the pipe. The habitat is then dewatered and welding is carried out under similar conditions to hyperbaric welding, except that the pressure is one atmosphere. The welding habitat is left on the bottom around the pipe when the weld is completed.

Personnel and Qualification

The question of whether divers should be trained to be welders or whether welders should be trained to be divers is often raised. The answer is that whichever route is chosen, the selection and qualification of personnel is difficult and expensive. For pipeline tie-ins and repairs at North Sea depths, COMEX uses a dry transfer technique where the bell is locked directly onto the habitat. This has the advantages of reducing transfer times for personnel and equipment, and the welders descend and work in "shirt-sleeve" conditions. Consequently, experienced surface pipeline welders and pipefitters who are not necessarily divers can be employed. Before this, they will, however, have passed some 3 or 4 training and welding saturation dives for familiarization and qualification. The rules applying to the qualification of hyperbaric welders are similar to those in force for surface welding. For example, changes in pipe size, welding position, process, etc. require qualification. In addition, the welder has to be qualified at least to that depth at which he will operate. In practice, that means that each welder will perform at least one qualification test per year. When it is considered that the cost of requalifying a welder in the hydrosphere is on the order of $30,000 a year and can be 3 or 4 times this figure for training and qualification of a new welder; that for physiological reasons they are not allowed more than 5 to 6 saturation dives per year; that a turnover of 20 percent is usual – then it can be seen that the investment in personnel is extremely high.

For shallower repairs where saturation diving is not required and wet transfers are used, then divers with surface welding experience can be trained and qualified. In fact, the present policy within COMEX is to train welders as divers and some divers as welders in order to produce operators with polyvalent skills employable over all types of worksites.

Examples of Repairs Undertaken

The method and procedure of repair to be adopted will obviously depend on the type of structure, location, depth, conditions, etc. Even for pipeline repairs, small changes in procedure will be required from one worksite to another. For shallow water repairs, a simple habitat and wet transfer can be used if the diving tables indicate sufficient time is available per dive for each stage of the work. For North Sea, seabed operations and saturation diving with full surface support are required.

Pipeline Tie-ins or Repair. To date, COMEX has built some six welding spreads for pipeline tie-ins and repairs. (A spread consists of pipe alignment capabilities and welding habitat.) These range from a massive 220-ton combined pipe alignment frame and habitat (Fig. 9), to a smaller 50-ton habitat with reduced alignment capabilities (Fig. 10), to a simple 12-ton habitat used in shallow water with wet transfer. The latter two habitats are used in conjunction with seabed lifting frames or H frames (Fig. 11).

Fig. 9 – 220-ton lifting and alignment frame used on ETPM 701 Barge – Frigg Field (1976)

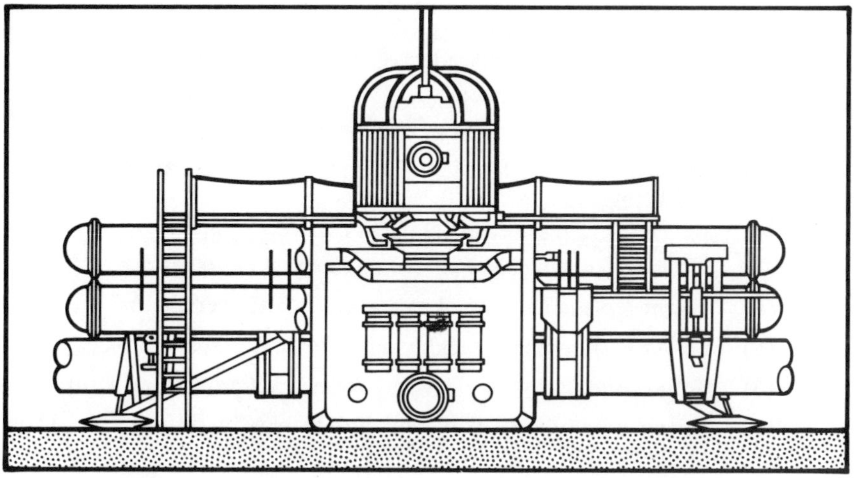

Fig. 10 – 60-ton alignment frame and welding habitat with dry transfer

Fig. 11 – H frame used for rough alignment

A typical sequence of events for removing and repairing a length of pipeline that has suffered external damage, corrosion, etc. (such that a length of pipe has to be cut out and replaced) would be as follows:

(1) Inspection dive and video of damage area; preparation of work procedure; mobilization of equipment

(2) Remove loose fill from pipeline; prepare landing areas for habitat and lifting frames

(3) Remove concrete (either by water jetting or diamond wheel); cut out length of damaged pipe and recover pipe to surface

(4) Lower replacement spool piece and cut off overlengths

(5) Lower H frames and roughly align flowline and spool piece

(6) Connect habitat guidelines and lower habitat over first joint

(7) Connect umbilical; place and inflate pigs in flowline and spool piece; fit doors; dewater habitat

(8) Final alignment; cut and bevel pipes; install pup piece, preheating equipment, etc.

(9) Weld side 1, weld side 2; x-ray welds and wrap pipe

(10) Cut away doors; lower pipe to seabed; recover habitat; reposition alignment frames on other end of flowline

(11) Repeat steps (5) to (10)

(12) Carry out pressure test on line and recommission

For a large diameter pipeline repair in the North Sea, the above operations would take some 20 to 30 working days. The actual pipefitting, welding, and inspection taking some 4 days. Requirements include a barge with saturation system, 35 to 45 people, and a vessel and crew at a daily rate of $60 to 130,000 depending on the vessel (40 to 60 percent of the cost). Personnel account for 20 to 35 percent of costs, and equipment and consumables, another 20 to 25 percent.

Pipeline Damage at Shallow Depths. Local repairs can be made where a section of line does not require being cut out, and effective repair can be made by patch welding (Fig. 12) or applying a mechanical joint and seal welding to the pipe. In this case, the diving system can be considerably reduced, and, since major pipe alignment is not required, the habitat can be a basic box with a minimum of support functions. In this case, a wet transfer would be used and GMA welding (either with heliox shielding gas if gas regeneration is available or argon/CO_2 if the welder/diver is to operate with his mask). A simple filtration system can be used to remove welding fumes. Using a supply boat with a team of 8 to 16 men should be enough.

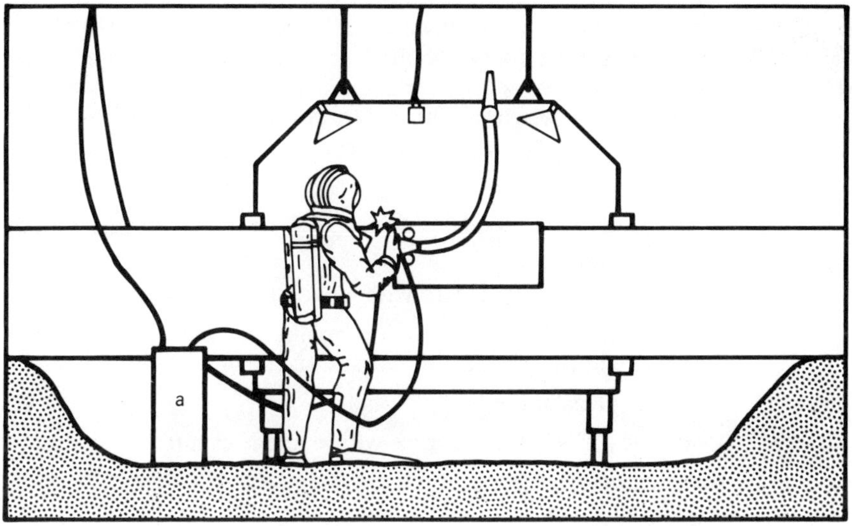

Fig. 12 – Patch welding in a simple habitat

Where a section of pipeline has been severely damaged such that a replacement spool piece has to be fitted and welded, it may still be carried out with a reduced spread using simple turnbuckles, pull-lifts, etc. within the habitat and a bias cut of up to 3° on the weld (Fig. 13). The bias cut is made using a hydraulic, side-cut milling machine. Down to 30 meters, the habitat gas can be compressed air; below that, for security reasons helium would be used. Again, a wet transfer with divers wearing individual breathing masks would be used with GMA welding.

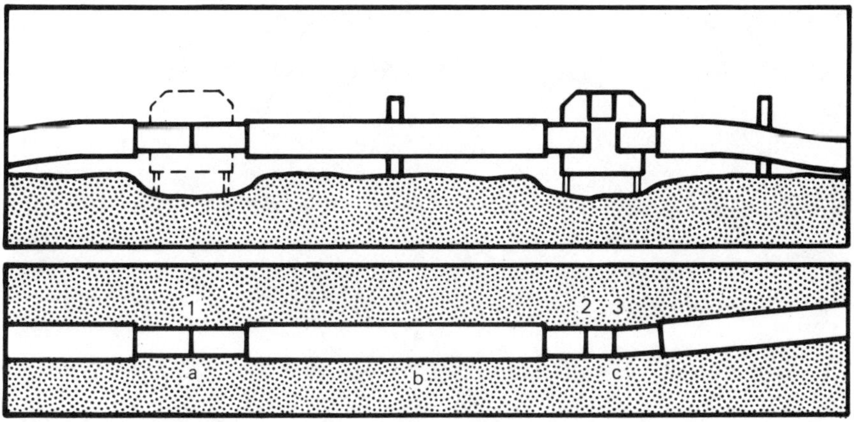

Fig. 13 – Principle of hyperbaric welding repair with one bias cut

The greatest incidence of damage to vertical pipes (risers, piles, etc.) is anticipated from the splash zone down to 15 meters depth. COMEX has developed a simple inflatable rubber habitat for cheaper repair of risers, etc. (Fig. 14). This can be affected either from the platform or a smaller support vessel and in some cases down to 30 meters without having to resort to saturation diving. The procedure consists of:

(1) Removal of the riser clamps

(2) Installation of a work platform, which is clamped around the riser

(3) Cutting off the upper length with a hydraulic saw (Fig. 15)

(4) Lowering the new length of riser with the weld bevel readily prepared and the habitat folded back on the pipe

(5) Installing plugs in the two pipe ends

(6) Centering and aligning the pipe with a simple clamp

(7) Inflating the chamber

(8) GMA welding of the joint

(9) Inspection of the joint, either by gammaradiography using a marinized source presealed films or by ultrasonics with surface readout

(10) Evacuation of worksite

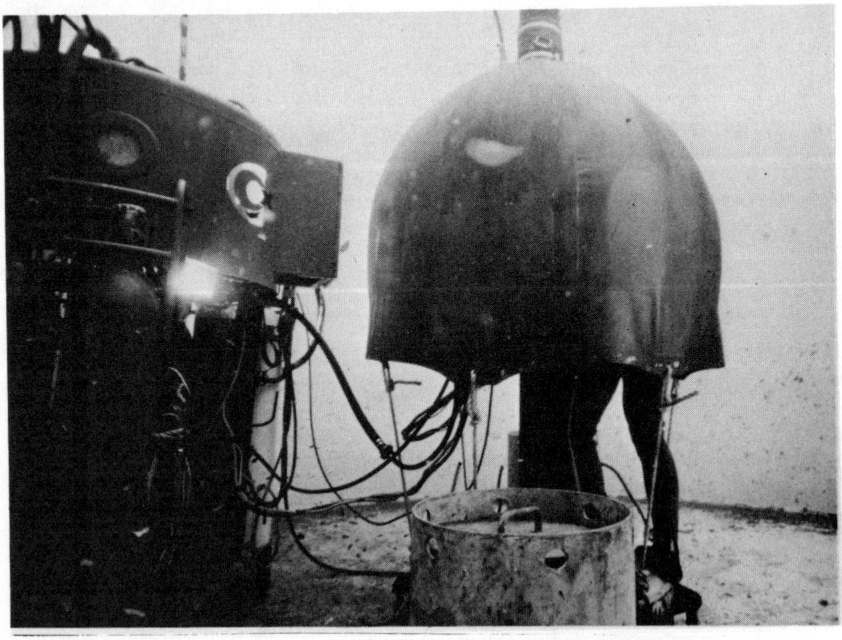

Fig. 14 – Rubber habitat for GMA welding

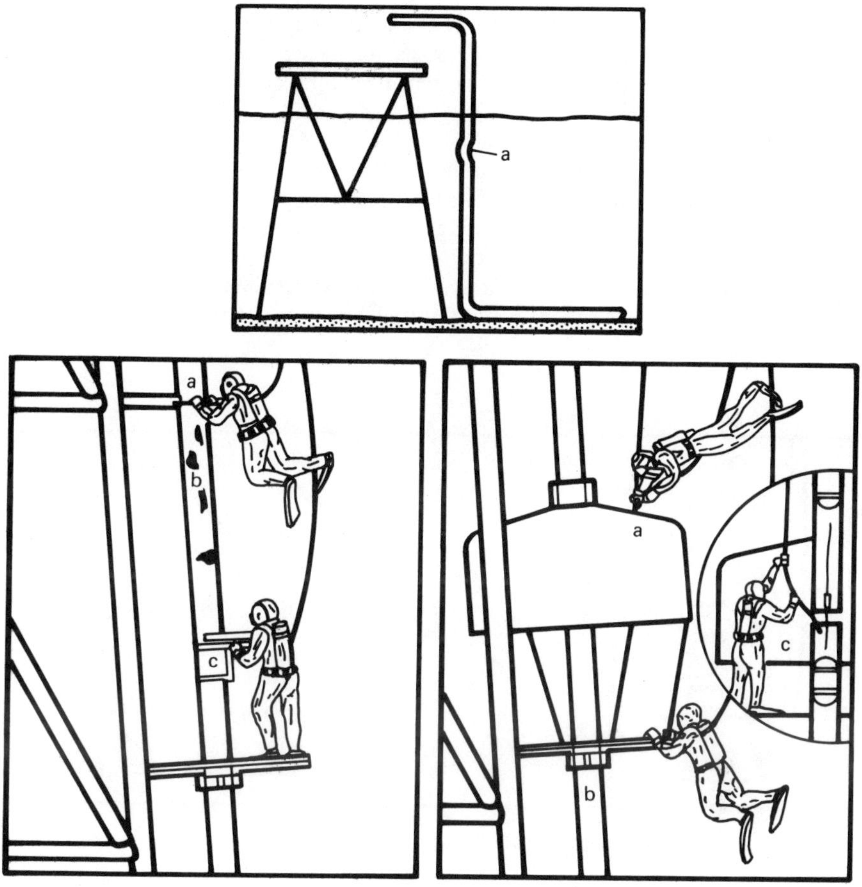

Fig. 15 – Different sequences of a riser repair using rubber habitat

Each stage of the work can be carried out in less than 3 hours including welding. For large diameters, two welders would be employed. This technique was successfully used at 25 meters depth for repair to a vertical structure in the North Sea where a reinforcement plate was welded in an area subject to fatigue.

A typical example of a habitat for node welding repairs is shown in Fig. 16. These are made in sections from measurements obtained from a survey dive and can be constructed on-site using steel and neoprene sealant. A jacket in Gabon was repaired at a depth of some 12 meters using this type of habitat. In this case, however, the floor was made an integral part of the chamber and the welders descended by a chimney to weld in the dry at atmospheric pressure. Similar chambers have been constructed for work at greater depths in hyperbaric conditions. COMEX's latest development consists of a rubber chamber surrounding the node that is sealed by a zipper. The advantages of this chamber are low cost and short installation time, using the same techniques as described for the riser repair. We expect to use this chamber for 11 node repairs in the Gulf of Suez on a drilling platform with fatigue cracks.

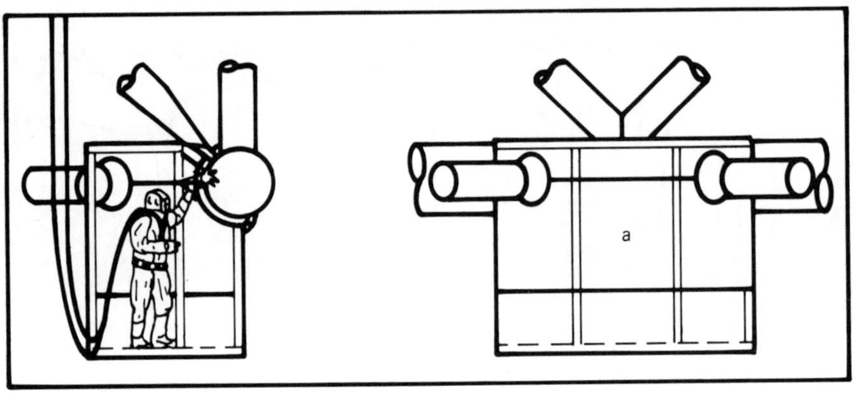

Fig. 16 – Dry welding repair of a node

On vertical wall repairs, hyperbaric welding can be effectively used, the chamber being clamped to the wall (Fig. 17). Fixing points at the moment have to be wet welded to the structure or by in-water stud welding. All subsequent operations are done in the dry. This type of chamber has been used to repair piling in a port development. A total length of 45 meters at depths from 0 down to 15 meters was welded. Subsequent ultrasonic inspection was used to examine the weld quality.

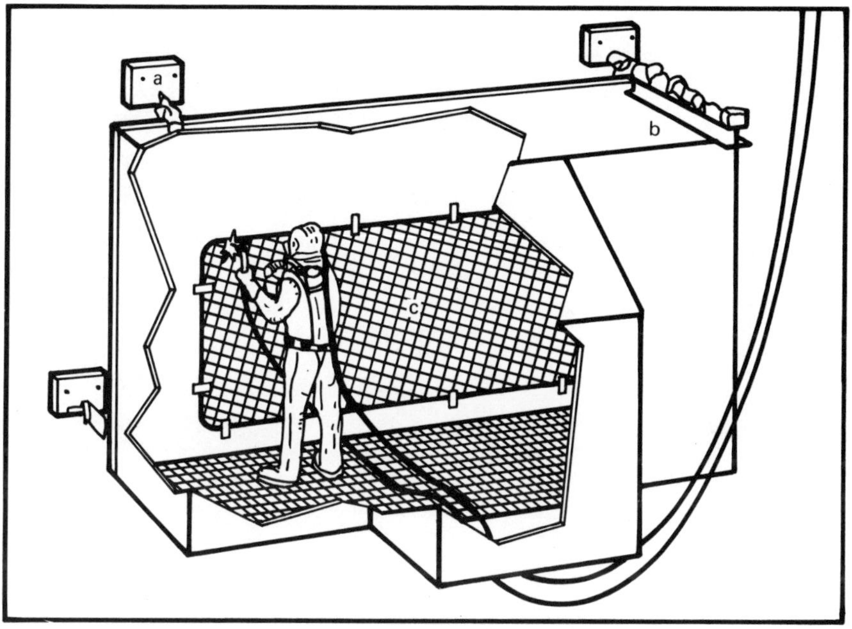

Fig. 17 – Dry welding on a vertical wall

Conclusions

Development of hyperbaric welding techniques has proved that welds can be obtained at 300 meters depth with properties equal to those obtained on surface. If we look at the growth of hyperbaric welding from 1975, it can be seen that this technique has now become established with the field of application growing each year.

At the present time, welding at 300 meters has been proved. However, it is certain that this depth can be increased to keep pace with the increase in working depth. Divers who were once considered as underwater laborers are becoming much more professional as technicians, fitters, inspectors, etc.

At the same time, welding costs have been reduced by the use of lighter equipment and vessels, and can now compete with the cost of mechanical tie-ins and repairs.

Future developments will probably include "wet" GMA welding, plasma cutting, and automatic pipe welding. Where saturation diving is no longer practical, developments in one-atmosphere welding and pipeline connection will be available.

Underwater Explosive Welding and Forming

David J. Leidel
Jet Research Center, Inc.

Introduction

During the last twenty years, explosive welding and forming have begun to be used as an alternative to conventional metalworking methods, particularly in applications where assembly or forming must be performed on materials or design configurations difficult to weld or form by conventional methods.

Although the destructive capability of explosives is far more widely applied in blasting, cutting, or demolition, explosives may provide a source of energy capable of performing useful nondestructive work, particularly in offshore operations. Two metalworking techniques are described, explosive welding and explosive forming.

This paper describes the basic principles of explosive welding and forming with examples of where these methods have been applied to solving offshore pipeline or platform construction problems.

Explosive Welding

That metals could be bonded together as a result of a high velocity collison has been known for many years. However, it was necessary to perform extensive theoretical and experimental studies to understand the mechanism of explosive welding and the geometrical and material variables that govern the process.

Explosive welding may be considered a process by which two or more metals are brought into sufficiently close contact so that the interatomic forces are strong enough to bond the metals together across the weld surface.

Since most metal surfaces are characterized by random or ordered surface asperities, oxide films, or surface contaminants, normal hydrostatic pressures are not sufficient to break through these surface conditions to achieve bonding. However, the pressure achieved by the detonation of a high explosive, although very short in duration, is high enough in magnitude to establish a bond interface between two metals. At present, over two hundred metals or combinations of metals are successfully bonded (Ref. 1). To describe the process of explosive welding, a simple parallel flat plate model was selected to describe the process and the variables that determine the success or failure of the method to obtain the desired bond. Figure 1 depicts a typical explosive welding configuration between two parallel plates, a base plate and a flyer or cladder plate. The base plate is generally supported by a thicker metal plate or anvil whose purpose will be described later. The pad of explosive is originally a layer of uniform thickness on top of the flyer plate, usually separated from the flyer plate by a rubber buffer. The process of explosive welding consists of four stages:

(1) Initiation of the high explosive and propagation of the detonation wave

(2) Acceleration of the flyer plate to its collision velocity

(3) Arrival of the flyer plate at the surface of the base plate and the formation of a re-entrant jet

(4) Formation of the weld; the bond interface generally characterized by a wavy interface

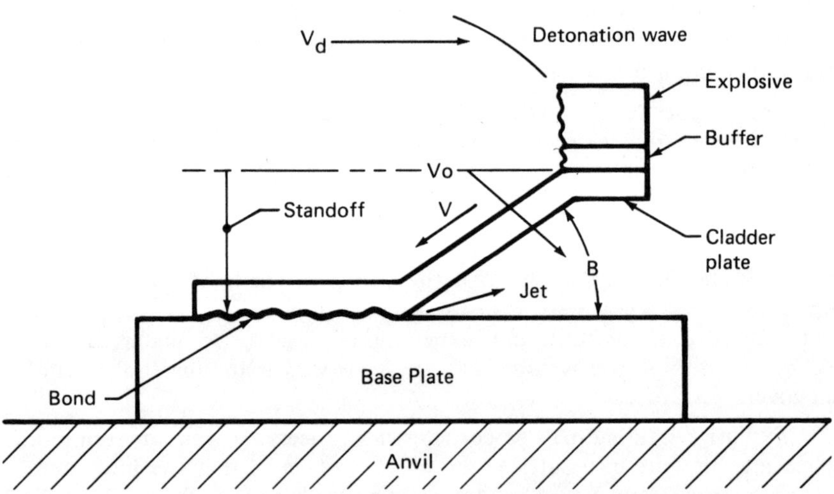

Fig. 1 – Parallel plate explosive welding configuration

The initiation of the explosive, generally by an electric blasting cap, results in a detonation proceeding along the explosive pad of grazing incidence to the flyer plate. The detonation wave, consisting of a shock wave and a thin reaction zone, ultimately becomes nearly constant in velocity, resulting in a steady pressure wave sweeping along the flyer plate. Although in grazing incidence the pressure imposed on a metal is approximately one-half the pressure produced by a normal detonation wave, the detonation pressure of the explosive may easily exceed 200 Kbar at the explosive/metal interface (Ref. 2). Although these pressures are of short duration, they are sufficient to accelerate the flyer plate to its collision velocity. The collision velocity of the flyer plate may be predicted by a formulation developed by Gurney, which relates to the velocity of a material propelled by an explosive to the mass of metal, mass of explosive, and a constant related to the energy release of the particular explosive (Ref. 3). For example, a typical Gurney equation predicting the collision velocity of the flyer plate for a pad of explosive propelling a flyer plate would be:

$$\frac{Vo}{\sqrt{2E}} = \left[\frac{(1 + 2\frac{M}{C})^3 + 1}{6(1 + \frac{M}{C})} + \frac{M}{C} \right]^{-1/2}$$

$$\sqrt{2E} = 2.93 \text{ mm}/\mu\text{sec for RDX}$$

where:

Vo is the collision velocity
M is the total metal mass
C is the total explosive mass

2E is the Gurney Constant; related to the energy release of the explosive and determined by terminal velocity experiment (Ref. 4).

A standoff distance is provided (somewhat exaggerated in Fig. 1), to allow sufficient distance for the flyer plate to reach the suitable collision velocity and to control the angle formed by the flyer plate and base plate. (β in Fig. 1.)

Following the motion of the flyer plate across the standoff region, a flyer plate element makes contact with the base plate or parent material at the collision zone, each flyer element colliding with the base plate and advancing the collision zone with the sweeping detonation wave. In the simple parallel plate model described, this collision zone advances at the same velocity as the velocity of detonation of the explosive. The phenomenon, which most investigators believe essential to the formation of a good weld, that must occur at the collision zone is the formation of a re-entrant jet (Ref. 5).

The formation of a small jet of metal (Fig. 1) is analogous to the jet and slug formation occurring in shaped charges or lined explosive cavities (Ref. 6). The re-entrant jet, a combination of cladding plate material and base plate material, removes the surface oxides by scarfing the top surface layer of both plates prior to the formation of the weld. The re-entrant consists of a very small portion of the cladder and base plate material, this mass fraction becoming much smaller with decreasing collision angle β.

The formation of the weld interface is generally characterized by a wavy appearance. Some investigators postulate that when collision bonds are formed, particularly at higher metal stream velocities where the flow is no longer laminar, flow separation occurs, resulting in cavity formation in the weld zone (Ref. 7). Kowalick and Hay propose that flows form in the cavities, these vortices ultimately being shed into a von Karman vortex street (Ref. 7). However, turbulent flow is not necessary for weld formation to occur; Keller et al. propose that a wavy but laminar interface is necessary for optimum weld quality (Ref. 1).

Variables Governing Weld Quality. The following are some of the important variables that govern the formation and quality of an explosively formed weld:

(1) Flyer collision velocity. The collision velocity of the flyer plate is governed by the explosive/metal mass ratio and the standoff distance as original spacing between the flyer and base plate. The flyer plate velocity must be sufficiently high to produce an impact pressure one order of magnitude higher than the dynamic yield strength of the strongest material to be welded. For iron this would be a pressure of approximately 60 to 80 Kbar. The impact velocity would then be determined from an impact analysis using the cladder and base plate material.

(2) Standoff distance. The initial standoff distance between the flyer and base plate permits the flyer plate to achieve its collision velocity. This velocity is reached after several reflections of the shock wave through the flyer plate thickness and requires a gap width of approximately one-half to one flyer plate thickness. It is absolutely essential that the standoff region be kept free of all liquids or contaminants. A dewatering operation is crucial to proper weld formation where liquids may be present, as in underwater operations.

(3) Collision angle β. If it is to be assumed that the formation of the re-entrant jet is necessary for a properly formed weld, then a governing condition exists limiting the collision angle β to a range larger than β critical. The relationship between collision velocity (Vo), collision angle (β), and detonation velocity of the explosive (V_d) for the parallel model is the Taylor Relation:

$$Vo = 2\,V_\alpha \sin\left(\frac{\beta}{2}\right)$$

The angle β must be adjusted to assure that the velocity of the cladding plate with respect to the moving collision points remains subsonic, or less than the bulk sound speed of the material to be welded. Chou, Carleone, and Karpp describe this criteria and the mechanism by which an attached shock at the collision point would result in a jetless collision for collision angles less than β critical (Ref. 8).

(4) Shock Propagation Effects. Since explosive welding utilizes high velocities and pressures applied to elastic-plastic solids, the interaction of the shock waves or pressure discontinuities in the materials becomes important. The nature of the shock wave interaction when two materials (either identical or dissimilar) collide is governed by the laws of conservation of mass, momentum, and energy and the equations of state or Hugoniot relations for the materials. An important quantity known as the shock impedance ($\rho_0 C$) is defined as the product of the density of the undisturbed material and the shock velocity. The relative magnitudes of the shock impedances of the cladder and base plate determine the magnitude and sense (tensile or compressive) of the incident, transmitted, and reflected stress waves propagating in the materials. If the recombinant wave fronts reflected from the boundaries of the base plate and flyer plate are strongly tensile, the recently formed bond may tear apart under the action of these stresses. This is a major factor determining which combination of metals may be successfully welded explosively. Ezra states that the ratio of the shock impedance of the base plate and flyer plate should not exceed approximately three (Ref. 1). Furthermore, an anvil or thick support plate should be used to stiffen the base plate and should also be the same material as the base plate to assist in delaying and attenuating stress waves reflected from the free surface.

(5) Explosive Detonation Velocity. Since the initial angle between the base plate and the flyer plate is generally either zero or a very small angle, only explosives with very low detonation velocity are suitable for explosive welding. In some early experiments, low velocity dynamites or nitroguanidine were commonly used. However, specially formulated welding powders that are cap-sensitive with predictable detonation properties are currently available (Ref. 9). Low velocity explosives ensure that the critical relative flyer plate velocity criterion will not be violated.

(6) Ductility. In general, increased material improves ductility. Noble metals such as silver are easiest to explosively weld. But copper, ductile steel, aluminum, and nickel have been determined to be amenable to explosive welding.

Properties of the Explosive Weld. Only a very limited amount of data exists regarding the strength and metallurgical properties of the explosive weld.

In steel specimens tests have shown that the material in the vicinity of the explosive bond shows a marked increase in hardness: approximately 40 points (Vickers) for the stainless steels and 80 points for 1020 hot-rolled steel (Ref. 10). Nonferrous alloys may increase in hardness by 100 percent (Ref. 1). Stress relieving prior to use may be necessary if the possibility exists of straining the bond zone beyond the ductile limits or loading the bond zone by impact loads.

Shear strength tests reported by Ezra noted that shear failure usually occurs through the flyer plate or cladder plate material and not through the bond zone (Ref. 1). Shear strength is essentially the same normal to and parallel with the wavy bond interface.

There is insufficient space in the body of this report to present the details of the material properties of the varieties of explosively clad composites. A suggested source for precise information on the mechanical properties of a specific case mav be obtained from the Deta Clad Division of E.I. DuPont de Nemours.

Explosive Welding of Offshore Pipelines. Two commercially available applications of explosive welding in offshore operations are the all-welded connection for subsea pipeline and the welded pipe-to-flange connection (Ref. 11). A typical configuration for an axisymmetric explosively welded connection is shown in Fig. 2, in this case an all-welded pipeline connection developed by Vickers Offshore. Initial commercial development and qualification tests were conducted on 16 in.-OD API 5LX60 line pipe.

To achieve a successful explosive weld on commercial line pipe, the following steps are required by the Vickers system:

(1) Machining of the outer diameter of the pipe to be welded to remove scale and ensure a minimum surface finish requirement.

(2) Insert explosive charge and dewatering bags. Dewatering bags reduce by several orders of magnitude the shock loading that could deform the pipe.

(3) Mechanically connect pipe sections, supporting pipe sleeve with a removable anvil.

(4) Inflate dewatering bags and dewater weld annulus. This operation is essential to proper weld formation.

(5) Detonate charge and inspect welded connection as required.

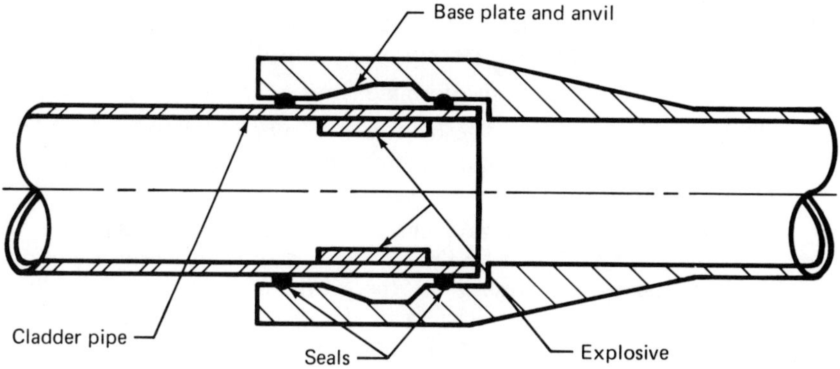

Fig. 2 – Explosively welded pipeline connection

A major factor noticed in this method of pipe assembly is the high degree of joint preparation completed prior to installation. A successful installation is less dependent on operator skill, as a diver need not be a qualified welder also. Although a system of this nature for underwater explosive welding is expensive, there may be applications in very deep water or unusually hazardous conditions where explosive welding may be cost effective. Questions remain concerning long-term fatigue life and corrosion resistance of this method of assembly, but, with continued interest by offshore engineers, research effort will be made to understand more fully the process of explosive welding.

Explosive Forming

In the previous discussion on explosive welding, it is noted that the explosive is essentially in direct contact or only slightly removed from the workpiece being welded; hence, explosive welding is termed a contact operation. Noncontact operations, where the explosive is located at some distance from the workpiece and the energy transmitted to the workpiece through a suitable medium (usually water), results in much lower pressures and resultant velocities and thus is used for explosively forming materials. The result is plastic deformation of the workpiece to the desired configuration without drastic alteration of the microstructure of the material or macroscopic mechanical properties of the material.

The general configuration for an explosively formed connection consists of an explosive charge, a liquid standoff distance, the workpiece, and a die or cavity into which the workpiece is to be formed.

The mechanism of explosive forming consists of the following events:

(1) Detonation of the explosive, usually a spherical, linear, or toroidal geometry, and the propagation into the water medium of the primary shock wave.

(2) The primary shock wave, containing approximately 50 percent of the energy released by the explosive, arrives at the surface of the workpiece and results in the workpiece acquiring an initial "kickoff" velocity. The velocity of the workpiece results in cavitation occurring in the liquid and subsequent unloading of the workpiece.

(3) The remainder of the energy released by the explosive is dissipated in an oscillating gas bubble which, in the process of pushing the liquid ahead of it, reloads the workpiece. Reference 1 presents in detail the various analytical techniques and experimental results of several configurations.

A typical analytical method of designing an explosively formed workpiece follows the following steps:

(1) Estimate the amount of strain energy required to form the workpiece to the desired configuration.

(2) Determine the amount of residual strain energy, if any, remaining in the workpiece die or support following deformation of the workpiece.

(3) Using empirically determined efficiency data, determine the explosive configuration required to supply the required energy. These data are available for some configurations if the energy transfer medium is water, since shock and energy transfer in this medium has been well documented.

This is a very general approach to designing a system to explosively form a particular workpiece. Reference 1 gives a number of configurations and the methods to calculate the design variables in each case. By far, the largest application of explosive forming has been in the aerospace industry in the forming of domes, combustion chamber liners, heat exchanger components, and aircraft components. However, a recent application of explosive forming in the offshore industry has been described by Dailey et al. (Ref. 13).

Application of Explosive Forming to Offshore Structures. Offshore engineers have expressed interest in recent years in the development of a mechanical pile-to-jacket connection in offshore structures such as production platforms. The conventional joining method to assist in the transfer of loads from the jacket legs of a platform to the piles has been the use of a grouting operation; grout being pumped into the annulus between the jacket and the pile. On occasion, the integrity of the grouted connection has been uncertain and is always difficult to inspect.

Figure 3 shows a mechanical connection consisting of a groove in the sleeve interior into which the pile is explosively swaged or expanded. This type of swagging operation may be accomplished by hydraulically cold working the pile wall into the groove. However, another method requiring much less expensive tooling utilizes a toroid-shaped explosive charge to expand the pile wall into the groove following the driving of the pile down the center of the jacket. The explosive system has the simplicity of no moving parts, is relatively light in weight, and is extremely cost effective in terms of energy output.

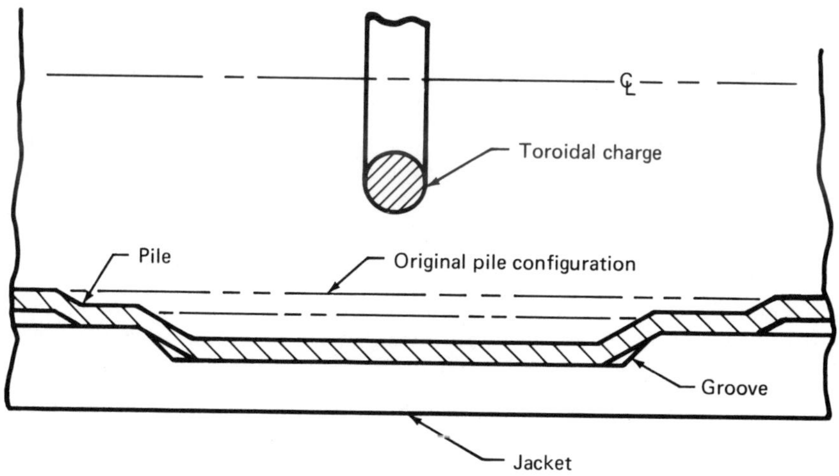

Fig. 3 – Explosively formed pile/jacket connection

Laboratory tests have been conducted on scaled pile/jacket connections (5-inch OD piles with 6-inch OD sleeves and 10-inch OD piles with 12-inch OD sleeves) to determine both static and dynamic axial load capacity (Ref. 13). The experimental results appear to agree well with the analytical model, and work is progressing to extend the tests to full-scale models.

Although precise cost analysis for the explosively formed sleeve-to-pile connection must be done for each specific case, preliminary studies indicate that the technique is economically attractive in comparison to conventional grouting.

Summary and Conclusions

The previous discussion is an overview of the methods by which constructive work may be done by a controlled detonation. Some methods are currently in common practice, while others are in the stage of research or development.

The requirement of brevity in the report limited discussion to only two methods of explosive utilization and two examples. Other explosive metal-working operations include explosive hardening of some ferrous alloys and explosive powder compaction. A more comprehensive discussion of the methods presented here may be found in the references listed. It is most useful to become familiar with these techniques to be able to appraise the potential of these novel construction methods in solving engineering problems offshore.

References

1. Ezra, A.A. 1973. *Principles and practice of explosive metalworking,* vol. 1, pp. 173-184. Great Britain: Garden City Press Ltd.
2. Pearson, J. 1972. The energetics of explosive operations. In *12th annual symposium on the behavior and utilization of explosives in engineering design,* pp. 69-84. ASME and Univ. of New Mexico.
3. Gurney, R.W. 1943. *The initial velocities of fragments from bombs, shells and grenades.* Ballistic Research Laboratories Report No. 405.
4. Kennedy, J.E. 1972. Explosive output for driving metal. In *12th annual symposium on the behavior and utilization of explosives in engineering design,* pp. 109-124. ASME and Univ. of New Mexico.
5. Cowan, G.R. and Holtzman, A.H. 1963. Flow configurations in colliding plates: explosive bonding. *Journal of Applied Physics* 34(4): 928, 932-933.
6. Pugh, E.M.; Eichelberger, R.J.; and Rostoker, N. 1952. Theory of jet formation by charges with lined conical cavities. *Journal of Applied Physics* 23(5): 532.
7. Kowalick, J.F. and Hay, D.R. 1971. A mechanism of explosive bonding. *Metallurgical Transactions* 2 (July): 1953-1958.
8. Chou, P.C.; Carleone, J.; and Karpp, R.R. 1979. Criteria for jet formation from impinging shells and plates. *Journal of Applied Physics* 47(7): 2975-2981.

9. Trojan Division, Commercial Solvents Corporation. *Technical data sheets.* Spanish Forks, Utah.

10. Rinehart, J.S. and Pearson, J. 1963. *Explosive working of metals,* p. 314. New York: Pergamon Press, MacMillan Co.

11. Vickers Offshore Underwater Pipeline Engineering Group. 1977. *Technical data sheet.* Barrow-in-Furness, U.K.

12. Cole, R.H. 1948. *Underwater explosions.* Princeton, N.J.: Princeton University Press.

13. Dailey, J.E.; Regalbuto, J.A.; Merwin, J.E.; and Sims, J.R. 1980. High energy formed connections, Paper no. 3787. In *Offshore Technology Conference.*

Underwater Factors Affecting Welding Metallurgy

Koichi Masubuchi
Massachusetts Institute of Technology

Abstract

This paper discusses fundamental aspects of underwater welding, especially "wet" arc welding processes. It covers factors affecting metallurgical properties of underwater welds. The subjects covered include (1) bubbles surrounding the welding arc, (2) heat flow in weldments and hardness, and (3) hydrogen-induced crack sensitivities of welds made in several steels.

Introduction

Published sources of information on the fundamentals of underwater welding are very limited compared to the complexity of the subject. Figure 1 shows a result of a survey conducted at M.I.T. a few years ago on papers on underwater welding published in various journals. Although underwater welding was first reported in the 1930's, there were only a limited number of papers on underwater welding until around 1960. Since then there has been a rapid increase of papers on underwater welding, as clearly shown in the figure. Although it is rather difficult to estimate the number of published articles prepared so far in the world, the estimate is that the total number would be a few hundred. This is far less than the total number of papers on welding, which may be estimated to be around 2,000 papers per year. In addition, most of the papers on underwater welding discuss practical aspects such as development of new processes. Therefore, the total number of papers ever published in the entire world on fundamentals of underwater welding is limited, probably less than 100. Publications from M.I.T. represent a significant portion of this small group, even though our research efforts so far have covered only a fraction of what is needed to be done.

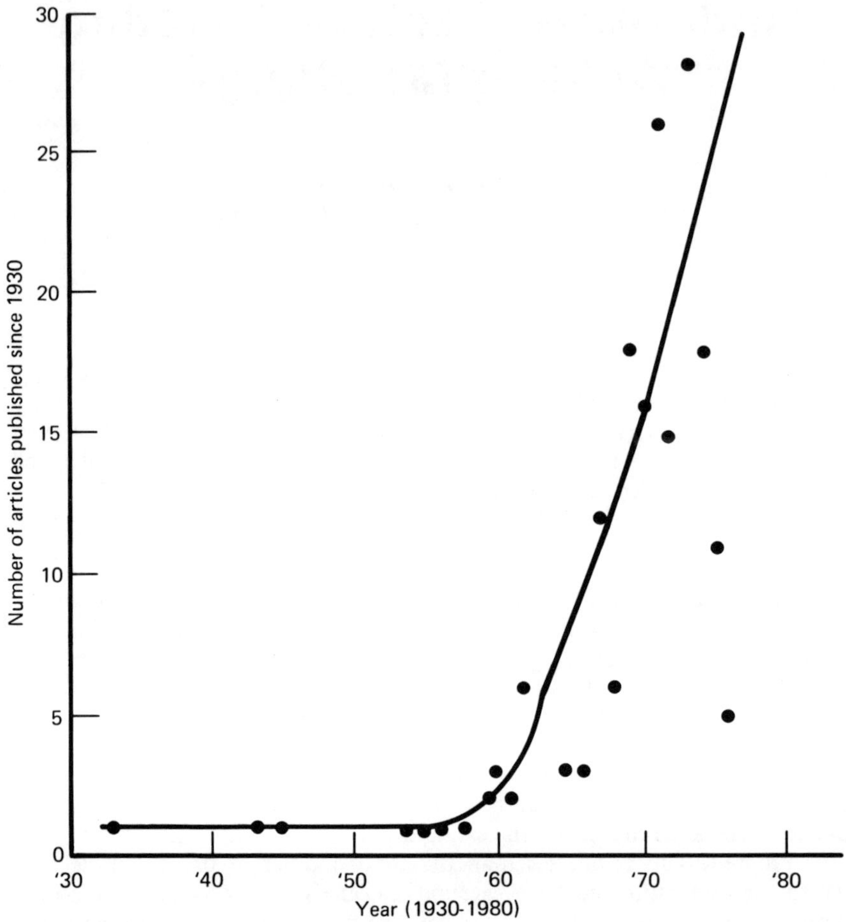

Fig. 1 – Articles on underwater welding published since 1930

Research on underwater welding started at M.I.T. in 1968. Although our research effort has been rather modest in terms of research expenditure, we have been able to produce 17 theses, including one for a Ph.D., fourteen for Engineer's and Master's, and two for Bachelor's degrees. Most of the students who prepared these theses are still active in various fields of ocean engineering and/or welding engineering. We also have published a number of papers and reports including two Sea Grant reports and a Welding Research Council Bulletin (Refs. 1-3). We hope to continue the effort in the future. Another hope is that more research on fundamental aspects of underwater welding will be conducted by many other investigators.

Underwater Welding Processes and Problems Involved

A number of welding processes have been used and studied for underwater applications. By far the most commonly used are arc welding processes including the shielded metal arc process using covered electrodes and the gas metal arc process. Other welding processes that have been used and/or studied include submerged arc, plasma arc, stud, thermit, friction, resistance welding, etc. Underwater welding processes can be classified into the following groups on the basis of how shielding is provided:

(1) Wet welding processes

(2) Dry welding processes including partially and/or locally dry welding processes and completely dry processes

The term "wet welding" is used to indicate that welding is performed underwater, directly exposed to the wet environment. In the simplest case of wet shielded metal arc welding, the arc is maintained between the electrode and the work, both of which are immersed in water. The welding arc is separated from the surrounding water by the gaseous environment composed of the gas produced from the covering and the gas formed by the decomposition of water due to the intense heat of the arc. Various methods have been developed to protect the arc from direct exposure to the surrounding water. In case of the flux shielded process developed at M.I.T., the arc is surrounded by flux (Ref. 2). In the case of underwater stud welding, most areas of the stud do not have direct contact with the surrounding water (Ref. 2). However, these processes should still be included in the "wet" welding processes, since the plates to be welded are completely exposed to the surrounding water. Major advantages of wet welding are its simplicity and its ability to be used in the wet environment. The major shortcoming is the rather poor quality compared to that of welds made in air. Underwater wet welding processes are widely used for repair jobs.

In the dry welding processes, regions of the metals to be joined are kept dry during welding. In the completely dry welding processes, welding takes place in the completely dry environment. When welding is performed under pressure higher than the atmospheric pressure, it is often called "hyperbaric" welding. Since welding is done in the completely dry environment, the quality of these welds can be as high as that of welds made in air. Therefore, the completely dry welding processes are widely used for critical jobs such as joining of underwater pipelines. The major problem of these processes is the extremely high cost, especially when they are done in deep sea.

In the partially dry processes, movable chambers are often used to create a dry environment in some regions near the metals to be joined. In most cases the body of the welder/diver is partially immersed in water. The partially dry processes have characteristics somewhat between the wet processes and the completely dry processes. The partially dry processes are less expensive than the completely dry processes. However, the quality of these welds is less than that of welds made by the completely dry processes. Even though the regions of the metals near the joint are kept dry, the atmosphere tends to have some moisture.

It is not the intention of this paper to present details of costs and qualities of welds made by these processes, since they are greatly affected by details of each specific process. As far as the basic mechanisms of welding are concerned, there is very little difference between ordinary welding in the air and dry underwater welding processes, except the effect of pressure in the case of hyperbaric welding. Consequently, the emphasis of this paper is placed on fundamental mechanisms of underwater wet welding, especially arc welding.

Obviously, the unique feature of the wet welding is that metals to be joined are immersed in water. Major effects of water on arc welding are as follows:

(1) Due to the quenching effect of water, a weldment cools rapidly, resulting in a hard and brittle weld.

(2) Bubbles are formed due to the intense heat of the welding arc and the weld metal may become porous.

(3) Hydrogen in the bubbles and the surrounding water may cause hydrogen-induced cracking.

The effects are further affected by the pressure as the depth increases and the salinity of water.

Even though only a limited number of papers have been published so far, these papers have covered a number of subjects important in underwater welding, as follows:

(1) Underwater bubble formation

(2) The underwater welding arc and metal transfer

(3) Heat transfer in underwater welds

(4) Effects of pressure on underwater welding

(5) Underwater welding polarity

(6) Waterproofing of electrode coating

(7) Microstructures and hardness of underwater welds

(8) Crack sensitivity of underwater welds

(9) Porosity and other types of weld defects

The M.I.T. Sea Grant reports and WRC Bulletin 224 present detailed discussions of these subjects (Refs. 1-3). This short paper covers only some of the subjects based primarily upon results obtained at M.I.T., namely (1) bubbles surrounding the welding arc, (2) temperature changes and hardness, and (3) crack sensitivity.

Bubbles Surrounding the Welding Arc. Since the arc energy is very intense, a portion of water surrounding the welding arc is vaporized and forms a relatively stable bubble void. The bubbles have some effects on welding as follows:

(1) The bubbles act as shielding between the surrounding water and the molten metals being transferred from the electrode and the weld metal. The presence of the bubbles reduces cooling rates of the weld metal and the heat-affected zone of a weldment made underwater.

(2) When the gas shielded metal arc process is used for underwater applications, the shielding gas supplied from the welding torch collides with the bubbles. As a result, any gas shielding technique that may be effective in welding in air tends to lose its effectiveness when it is used underwater.

(3) The gases in the bubbles can be major sources of porosity in the weld metal.

(4) Because of the high temperature of the arc, some hydrogen in the bubbles may exist as atomic hydrogen, which may diffuse into the weld metal and the heat-affected zone and may cause cracking.

Underwater Bubble Dynamics. Researchers at M.I.T. used high-speed cinematography to study mechanisms of generation of bubbles due to the high concentration of heat in the arc, their departures from the arc region, and final collapse during the underwater arc welding processes in the wet condition. As an example, Fig. 2 shows an underwater welding arc between a covered electrode in the downhand position and a steel plate. M.I.T. researchers also conducted analytical studies of the mechanisms of bubble formation.

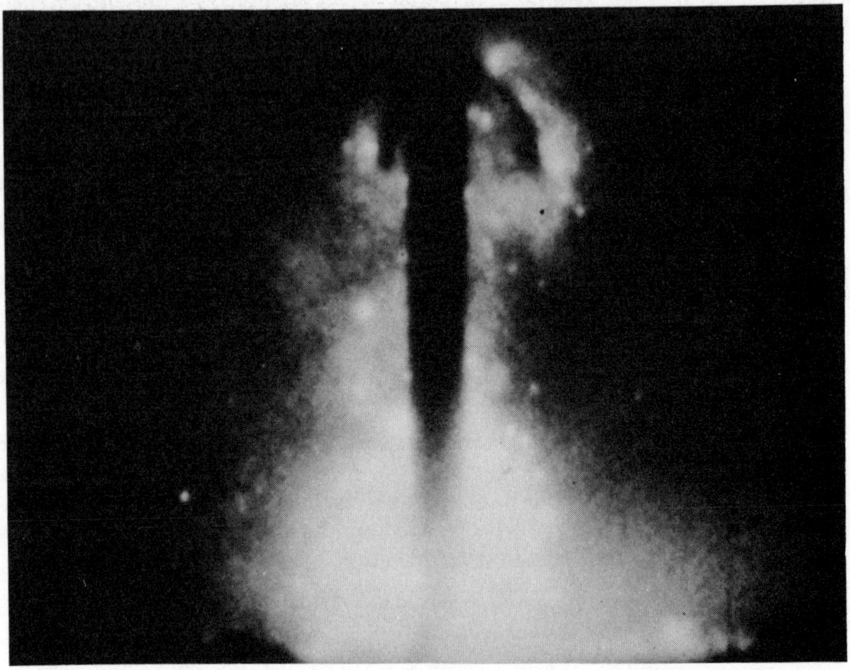

Fig. 2 – Underwater arc-bubble growth

Gases decomposed from the flux coating and vaporized from the surrounding water cause the bubbles to grow. A bubble grows and rises continuously until its radius becomes tangent to the initial void. The welding arc is always protected by the bubble during welding. The shape of the bubble, especially in later stages of its growth, may be assumed to be spherical. To predict bubble phenomena mathematically, an idealized bubble growth model with bubble formation from an orifice can be simulated. It assumes that the bubble growth begins at a point on or just above the plate and that the gas behaves ideally. Only the balance of buoyant and inertial forces is considered in this analysis. Large flow rates make it possible to neglect surface tension. Figure 3 shows the idealized bubble growth models used in the analysis. Details of the cinematographic and analytical studies have been presented in several publications (Refs. 1, 4, 5). This paper presents a summary of important findings.

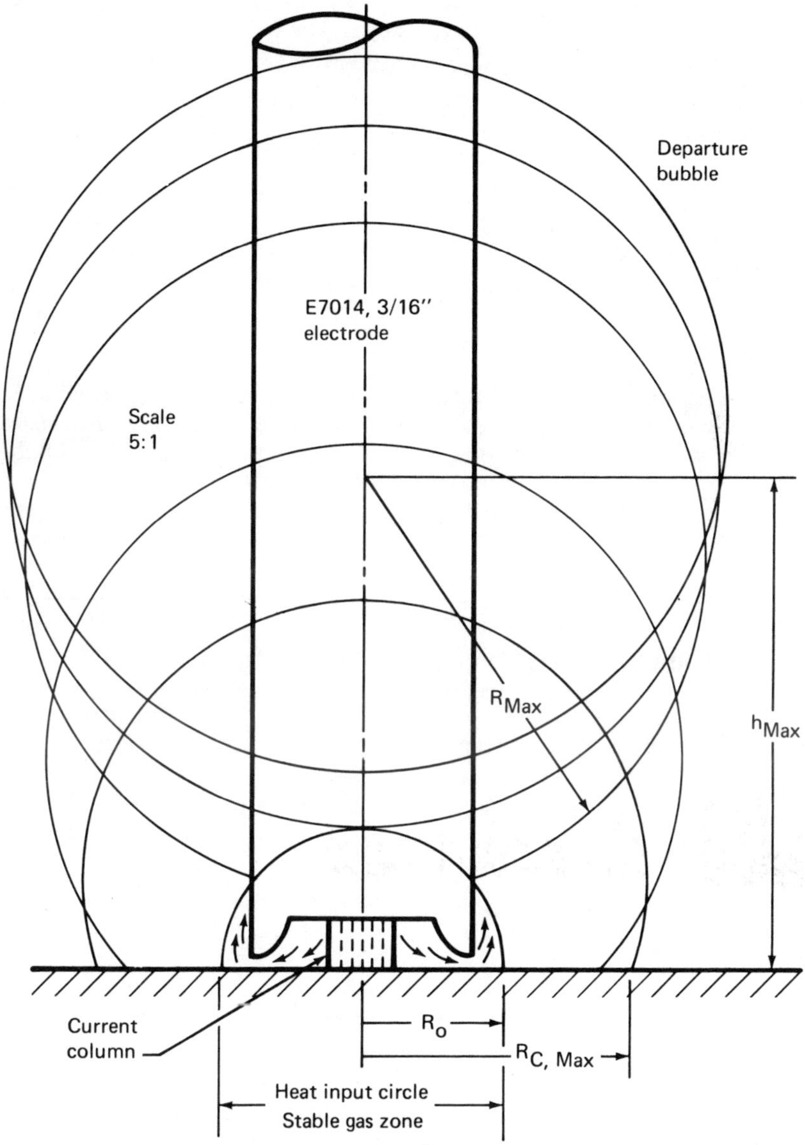

Fig. 3 – Idealized underwater arc-bubble growth

Breakaway of a bubble occurs when the bubble becomes tangent to the void or when its height equals its radius plus the radius of the void, R_o. The height of a departing bubble, h_{max}, can be written as:

$$h_{max} = R_{max} + R_o$$

The volume flow rate of the bubbles, G, can be expressed as follows:

$$G = \frac{4\pi}{3} R^3_{max} f_{bubble}$$

where f_{bubble} = bubble frequency

The maximum bubble radius, R_{max}, and the radius of stable void, R_o, can be measured by a high-speed photographic technique. The analysis can determine several values including (1) the maximum height of the bubble, (2) the period of a dynamic bubble (bubble frequency), (3) volume flow rate, etc. Table 1 compares measured and calculated values of several parameters that characterize the bubble formation phenomenon. A good agreement is obtained between the measured and calculated values.

Table 1
Bubble growth characteristics (E7014 electrode)

	Measured*	Calculated
R_o	0.5 cm	0.5 cm (measured)
Period	0.076 sec	0.071 sec
Frequency	13 bubbles/sec	14 bubbles/sec
G	60 cm^3/sec** (= 134 cm^3/sec) by calculation	134 cm^3/sec
R_{max}	1.31 cm	1.31 cm
h_{max}	1.83 cm	1.81 cm

* Measurement was done by a high-speed motion picture study conducted by A.J. Brown at M.I.T. in 1973.
**Measurement was done by collection of gases that have filtered through the liquid. Water vapor generated at the arc was eventually condensed, leaving only hydrogen and organic byproducts filtered through the liquid.

The total amount of gas generated has been reported by several investigators. Table 1 shows that the measured value of the gas flow rate is 60 cc/sec for E7014 electrodes. Brown reported that E6013 electrodes generated 40 cc/sec of bubbles (Ref. 6). Silva reported a rate of 50 cc/sec for E6027 electrodes and 60 cc/sec for E7024 electrodes (Ref. 7). Madatov reported rates of 33 cc/sec and 100 cc/sec for the two electrodes used (Ref. 8). While different electrode coatings produce different amounts of gases, the rate of gas generation for a certain type electrode appears to be little affected by changes of welding conditions. The results obtained so far suggest that the dissociation of water is primarily affected by the arc temperature.

Chemical Compositions of the Bubbles. A major portion of the gas bubbles results from dissociation of water by the extreme heat of the arc. This produces a large amount of hydrogen and oxygen. Most of the oxygen reacts quickly with the combustible elements of the electrode coating and molten metals to produce CO and CO_2. A small portion of the gas consists of metal vapor and various mineral salts from the electrode covering. Silva reported the following compositions (Ref. 7):

H_2 : 62-82%

CO : 11-24%

CO_2 : 4-6%

The remainder (approximately 3 percent) was composed of N_2 and metallic and mineral vapors.

Table 2 shows results of experiments conducted in actual diving conditions performed on the bottom of the Baltic Sea (Ref. 2). Gases generated by the welding arc were collected in a glass container (500 mL) and the compositions were determined by a mass-spectroscopic analysis. The container remained virtually gastight until the analyses were done, as indicated by the lack of oxygen. The relatively low values of hydrogen indicate that some hydrogen might have escaped during the collection. The types of electrodes used were manufactured in the Federal Republic of Germany.

Table 2

Chemical compositions examined by mass-spectroscopic analyzer of bubbles collected during welding in the Baltic Sea

Electrode type	H_2	CO	CO_2	Other
Rutile iron powder (E7014)	45%	43%	8%	4%
Titanium	30%	55%	10%	5%

These gaseous products of combustion of the weld metal, the parent metal, the components of the electrode covering, and dissociation of water, as well as the water vapor, continuously interact within the bubble environment. The degree of dissociation of water vapor is maximum at the interior of the bubble nearest the extreme temperatures of the arc and decreases rapidly toward the bubble-water interface.

Temperature Changes and Hardness. Due to the quenching effect of water, a weldment made underwater cools more rapidly than one made in air, resulting in a harder heat-affected zone and less ductile welds.

Heat Flow. Figure 4 shows schematically temperature changes of a point near the weld. Shown here are several curves as follows:

(1) Curve W, which shows actual temperature changes of a weld made underwater

(2) Curve A, which shows actual temperature changes of a weld made in air

(3) Curve T-1, which shows theoretical temperature changes based on the assumption that water always maintains direct contact with the weldment (neglecting the presence of bubbles)

(4) Curve T-2, which shows temperature changes using an improved mathematical model

Obviously, Curve W cools much faster than Curve A. However, Curve W cools considerably more slowly than Curve T-1. The major reason for the difference is that in actual welding the welding arc is surrounded by gas bubbles, which separate the weld metal from the surrounding water. Consequently, in order to develop a mathematical analysis of heat flow during underwater welding, one must consider among other effects the effect on heat flow of the bubbles near the arc. Tsai has developed a comprehensive mathematical analysis of heat flow during underwater welding considering many important factors involved (Ref. 9). Computers are used extensively for carrying out numerical computations. Since the analysis is very complex, details are not discussed in this paper. Tsai has succeeded in developing a mathematical analysis that gives reasonably accurate predictions. He also conducted analytical and experimental studies of the effects on cooling rates of various welding variables including welding arc power, welding speed, plate thickness, water temperature, etc.

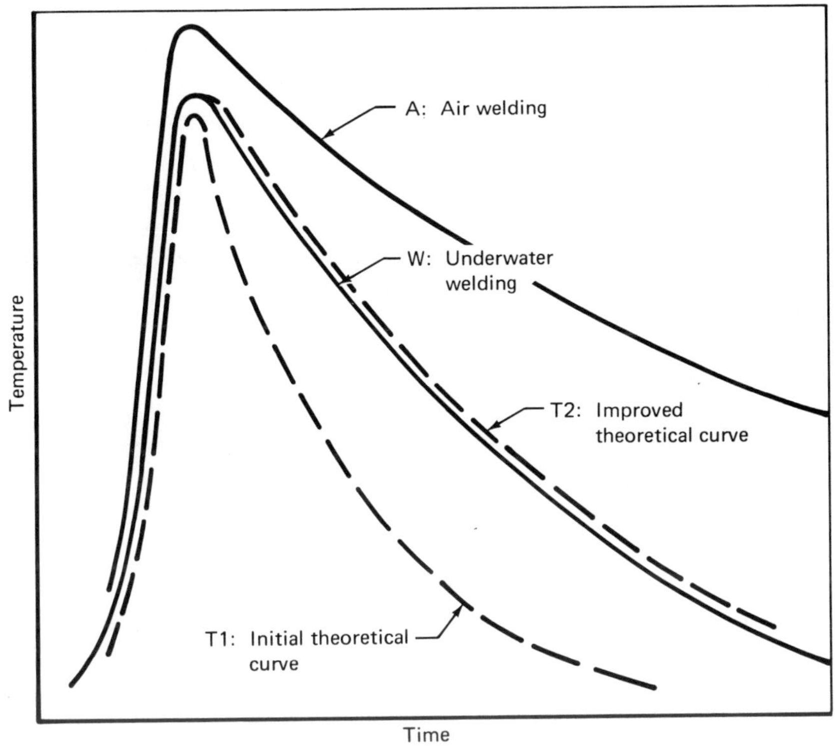

Fig. 4 – Schematical temperature changes during underwater and air welding

Methods of Reducing Cooling Rates. A number of investigators have studied various methods of reducing cooling rates of underwater weldments. Methods often used include (1) selecting proper welding conditions, (2) providing local dry spots near the arc, and (3) shielding some portions of a weldment from water. Figure 5 shows some results. Shown here are theoretical cooling curves at a point 1 mm from the fusion zone of welds made under the following conditions (Ref. 5):

Plate thickness: 1/8 inch
Arc power: 5,000 joules/sec
Water temperature: 76° F
Welding speed: 9 in./min

The cooling curves are calculated under the following conditions:

(1) Air welding.

(2) "Wet" welding with no insulation.

(3) Flux shielded process similar to submerged arc welding. The weld metal and the arc are covered with flux.

(4) "Wet" welding using a water jet that provides a dry spot surrounding the welding arc.

The cooling curves are superimposed on a continuous cooling transformation (CCT) diagram for 0.17% carbon steel.

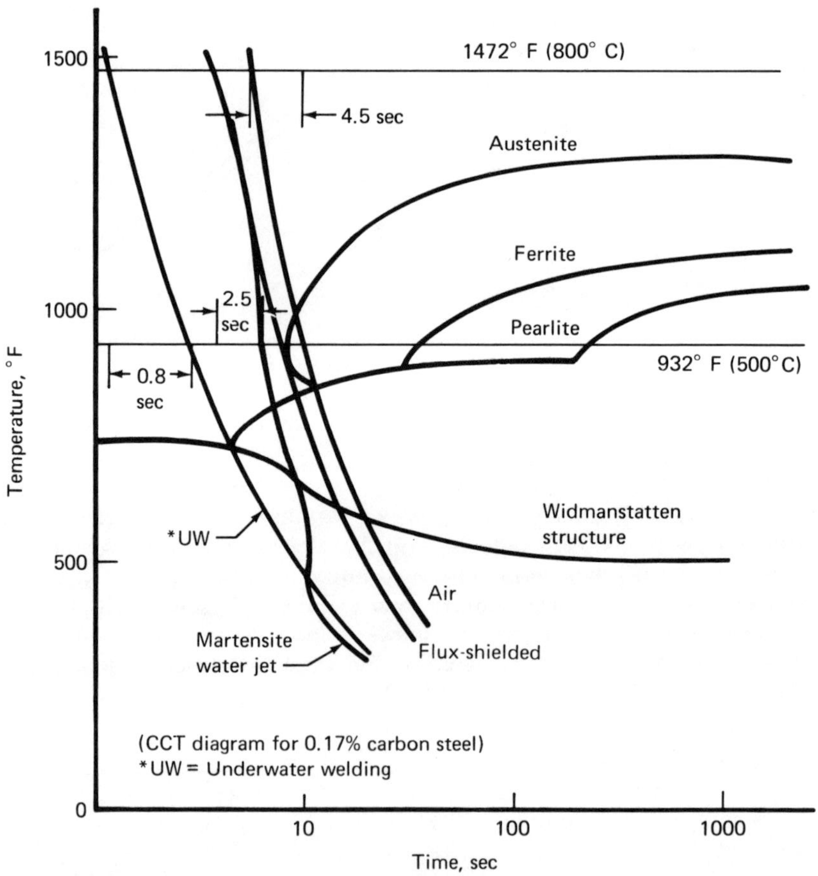

Fig. 5 – Effect of various insulation methods on cooling rate in heat-affected zone in UW welding

Obviously, the wet welding with no insulation produces the shortest cooling time. The figure indicates that the welds are transformed to martensite. The air weld provides the slowest cooling rate and the formation of martensite is completely avoided. The local dry spot technique is effective in reducing the cooling rate during a period immediately after welding and the point being considered is still within the dry spot. However, the cooling rate increases as the welding arc moves and the point being considered is exposed to the water jet. The flux shielding technique is very effective in reducing the cooling rate because the arc and the weld metal are covered by flux. The results suggest that the most effective method for reducing the cooling rate is to shield the welding arc and the weld metal from the surrounding water.

Hardness. Figure 5 indicates that martensite is formed when the cooling rate exceeds a certain critical value for the steel, which is called the critical cooling rate. As the amount of alloying elements increases and the strength level of the steel increases, the critical cooling rate becomes shorter and the hardness of martensite formed also increases. Figure 6 shows relationships between ultimate tensile strengths of steels and maximum hardness values of welds made underwater and in air (Ref. 2).

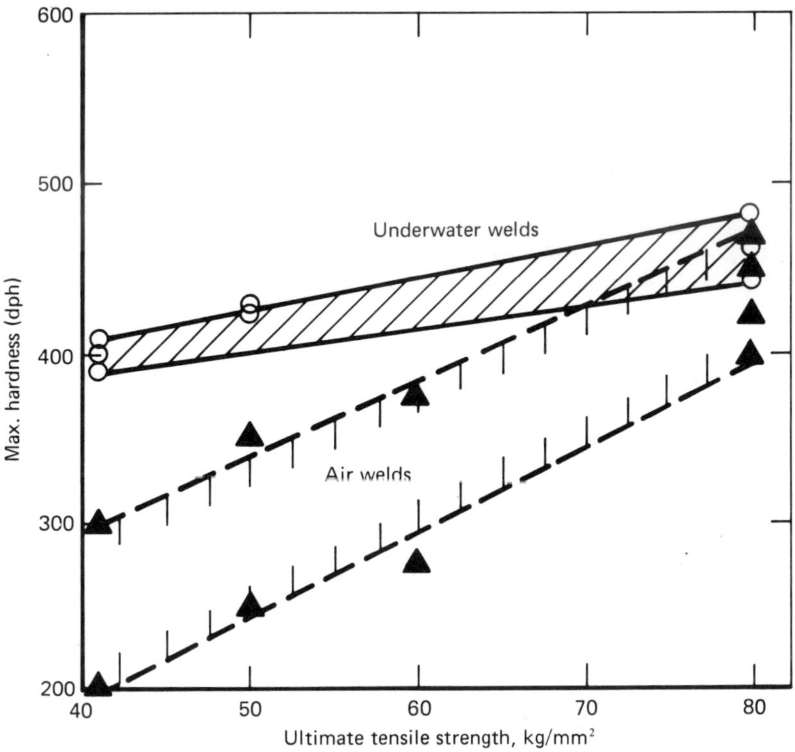

Fig. 6 – Relationship between maximum hardness of underwater welds and ultimate tensile strength of steels

As the strength level of steel increases, hardness values of both underwater and air welds increase. For steels with lower strengths, underwater welds are considerably harder than air welds. This is because martensite is formed in the underwater welds but none or only a limited amount of martensite is formed in the air welds. For higher strength steels the difference between the underwater and air welds is less, because hard martensite is formed in both welds.

Crack Sensitivity. Compared to welds made in air, welds made underwater have a higher tendency toward cracking, because the welds are quenched more rapidly and there is much more hydrogen. The tendency toward cracking is higher as the carbon equivalent values of the steels increase and when welds are under higher restraints. Although a wealth of published information is available on air welds made in a variety of steels, as a result of extensive research done over the years by numerous investigators in the world, published information on crack sensitivity of underwater welds is scarce. In fact, the information available is far less than what is needed to cover this complex subject affected by many design, material, and fabrication variables.

When fabricators develop procedures for welding certain joints made in certain materials, they are primarily interested in developing techniques for welding the specific joints successfully without cracking. They may publish some details of the successful techniques; however, they almost never publish data on cracking when unsuccessful techniques are used. Their working conditions (contractual and time constraints) almost never permit a systematic investigation of crack sensitivity of underwater welds. What complicates the matter is that crack sensitivity is determined by many design, material, and fabrication variables including details of joint design, plate thickness, alloying elements of the base and filler metals, welding processes, types of electrode coatings, welding conditions, etc. Even a joint cracks when certain welding techniques are used. The joint can be successfully welded by selecting a proper combination of welding variables. In fact, many practicing engineers are frequently trying to find out an optimum combination of fabrication variables for certain joints. Since an increasingly large engineering community is getting interested in underwater, it is extremely important to obtain information on general characteristics of crack sensitivity of underwater welds under various test conditions. In this short paper only limited data obtained at M.I.T. are presented.

Crack Sensitivity Data Obtained at M.I.T. A study was made at M.I.T. to investigate hydrogen-induced crack sensitivity of some underwater welds made in several steels. Since the results are published in the *Welding Journal*, only a brief summary is presented here (Ref. 10).

Used in the experiment are four commercially available steels including mild steel (ABS Grade A), two low alloy high-strength steels with 50 ksi yield strength (A537A and ST 52), and HY-80 steel, a low alloy quenched and tempered steel with 80 ksi yield strength. Table 3 shows chemical compositions of these steels.

Table 3
Chemical composition of materials used, wt-%

	Mild steel	ST52	A537A	HY-80
C	0.20	0.19	0.16	0.18
Si	0.02	0.34	0.30	0.20
Mn	0.53	1.14	1.20	0.32
P	0.03	0.02	0.02	0.02
S	0.04	0.03	0.03	0.03
Ni			0.20	2.99
Cr			0.20	1.68
Mo			0.06	0.41
V			0.06	
Nb			0.02	
CE[a]	0.29	0.38	0.45	0.85
P_{CM}[b]	0.23	0.26	0.26	0.36

a. Carbon equivalent $= C + \frac{Mn}{6} + \frac{Cr + Mo + V}{5} + \frac{Ni + Cu}{15}$

b. $P_{CM} = C + \frac{Si}{30} + \frac{Mn}{20} + \frac{Ni}{60} + \frac{Cr}{20} + \frac{Mo}{15} + \frac{V}{10} + \frac{Cu}{20} + 5B$

For the experiments on mild steel, three types of electrodes were used. They include titania-iron powder type (E7014), iron powder-iron oxide type (E6027), and low hydrogen type (E7018). The titania-iron powder type electrodes are known to have good running characteristics. The iron powder-iron oxide type electrodes have been demonstrated to reduce the hydrogen-cracking susceptibility of underwater welds.

Low hydrogen electrodes (E8018) were used for A537A steel and ST52 steel. For HY-80 steel, three types of electrode including low hydrogen electrodes with different strengths (E11018 and E7018) and austenitic electrodes (E310-16 [25Cr-20Ni]) were used. E7018 electrodes and E310-16 electrodes were used to determine exactly how an undermatched electrode and an austenitic electrode affect the hydrogen-cracking susceptibility of underwater welds in HY-80 steel. All of these electrodes were 5/32 in. (4 mm) in diameter.

Figure 7 shows the Y-slit type restrained cracking specimen used. The restraint intensity involved in this test corresponds to the upper limit of the restraint intensities of most welded joints used in actual construction. The measure of hydrogen-cracking susceptibility used was the cracking ratio as shown in Fig. 8. This is the ratio of the height from the root to the tip of the crack to the height from the root to the surface of the weld metal.

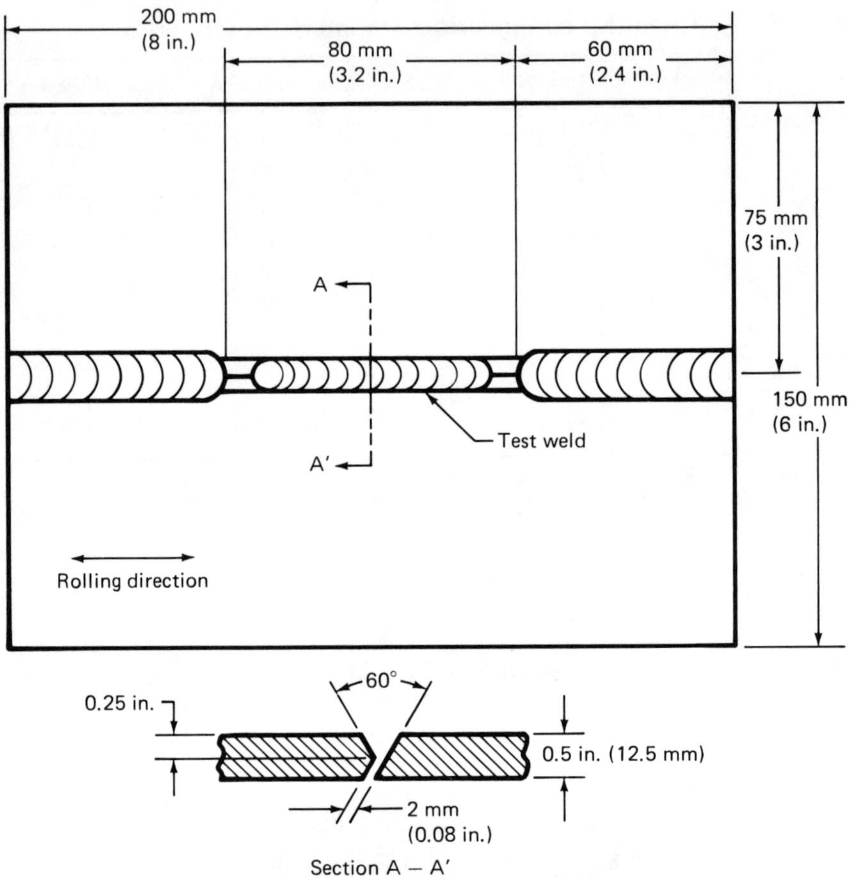

Fig. 7 – The Y-slit type test specimen

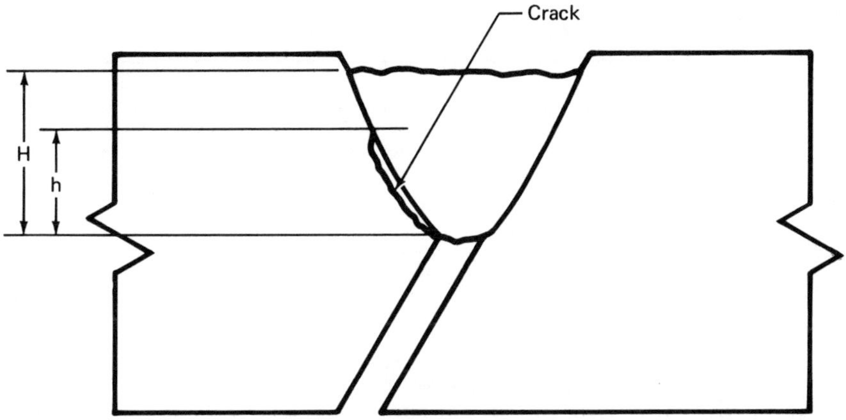

$$= \frac{\Sigma h}{\Sigma H} \times 100$$

Fig. 8 – Determination of a cracking ratio

Table 4 summarizes the results of the cracking tests, which can be stated as follows:

(1) No observable hydrogen-induced cracks were found in either underwater or air welds in mild steel. The electrode coating type did not noticeably affect the weld integrity.

(2) Underwater welds in high-strength steels cracked. The cracking ratios of underwater welds made in ST52 and A537A steels were 30 percent and 26 percent, respectively. The cracking ratios of air welds, however, were considerably different: 2 percent for ST52 steel and 21 percent for A537A steel.

(3) Underwater welds in HY-80 steel cracked extensively.

(4) The cracking ratio of HY-80 air welds decreased considerably by use of undermatched electrodes: from 95 percent for E11018 electrodes to 18 percent for E7018 electrodes. However, the cracking ratios of underwater welds were 100 percent for both E11018 and E7018 electrodes.

(5) The use of austenitic stainless steel electrodes reduced hydrogen-induced cracking. However, it resulted in an increase in hot cracking.

The paper published in the *Welding Journal* (Ref. 10) presents results of microstructures of welds, hardness, and amounts of hydrogen in welds.

Table 4
Summary of the cracking tests

Steel	Electrode	Cracking ratio, percent Air	Underwater
Mild steel	E7014	0	0
	E7018	0	0
	E6024	0	0
ST52	E8018	2	30
A537A	E8018	21	26
HY-80	E11018	95	100
	E7018	18	100
	E310-16	1	80

References

1. Brown, A.J.; Brown, R.T.; Tsai, C.L.; and Masubuchi, K. 1974. Report on *Fundamental research on underwater welding,* Report no. MITSG 74-29. M.I.T. Sea Grant Program.
2. Tsai, C.L.; Ozaki, H.; Moore, A.P.; Zanca, L.M.; Prasad, S.; and Masubuchi, K. 1977. *Development of new, improved techniques for underwater welding,* Report no. MITSG 77-9. M.I.T. Sea Grant Program.
3. Tsai, C.L., and Masubuchi, K. 1977. *Interpretive report on underwater welding,* WRC Bulletin 224. Welding Research Council.
4. Tsai, C.L., and Masubuchi, K. 1979. Mechanisms of rapid cooling in underwater welding. *Applied Ocean Research* 1(2): 99-110.
5. Tsai, C.L., and Masubuchi, K. 1979. Mechanisms of rapid cooling and their design considerations in underwater welding. In *OTC 3469, Offshore Technology Conference,* pp. 973-980.
6. Brown, A.J. 1971. *Methods of research in underwater welding.* B.S. Thesis, M.I.T.
7. Silva, E.A. 1971. Gas production and turbidity during underwater shielded metal-arc welding with iron powder electrodes. *Naval Engineer's Journal* December.
8. Madatov, N.M. 1965. The properties of the bubble of steam and gas around the arc in underwater welding. *Automatic Welding* 18(12): 25-29.
9. Tsai, C.L. 1977. *Parametric study on cooling phenomena in underwater welding.* Ph.D. Thesis, M.I.T.
10. Ozaki, H.; Naiman, J,; and Masubuchi, K. 1977. A study of hydrogen cracking in underwater steel welds. *Welding Journal* 56(8) pp. 231-s to 237-s.

Recently Developed Underwater Nondestructive Testing Systems

David B. Wyman and John Mittleman
Diving and Salvage Department,
Naval Coastal Systems Center

Introduction

During the past three years the Naval Coastal Systems Center (NCSC) has been developing and testing systems for conducting nondestructive examination (NDE) underwater on ships and fixed structures for the US Navy under the direction of the office of Ocean Engineering, Naval Sea Systems Command. The work to date has been centered around ultrasonic thickness gauging, magnetic particle inspections, and stereophotography. A philosophy of providing NDE work that is equal to or better in quality than is possible in dry dock has guided the development of these systems. In practice the philosophy has encouraged the use of equipment and procedures that do not count heavily on the NDE skills of the diver. Hard copy results are another important facet of the systems, for they greatly increase the confidence level of nondiving decision makers. In working with this philosophy, systems have been developed that consist of teams of Navy divers and qualified nondiving NDT technicians topside. The ultrasonic thickness gauging system makes the most use of this philosophy and is nearing completion of its development. The ultrasonic system has been tested on many thin hull inspections and is in the process of being accepted for use by the Fleet.

The stereophotography system provides a method of taking quality photographs, regardless of the underwater visibility and photographic experience of the diver. The photos are taken in pairs that allow stereographic viewing of them for engineering evaluation. The magnetic particle system is being developed into an easily transportable suitcase-size unit that will allow an MPI technician to go to a site and do a magnetic particle inspection using available Navy divers. Ultrasonic and electromagnetic flaw detection systems are presently under development.

Magnetic Particle Inspection

A magnetic particle inspection (MPI) system has been under development for the past few years. Initially, the work consisted of gathering background information on underwater MPI, principally work done in the North Sea. Next, various magnetizing systems and magnetic indicators were tested. Permanent, alternate current, and direct current magnets were tested including electroshock protection for the electromagnets. Various types of magnetic particles were tested as well as magnetic rubber, magnetic paint, and magnetic tape. Black lights for use in showing up fluorescent particles were developed and tested. Some additional hardware was also developed to make the system more easily worked underwater. After laboratory investigation of the hardware, a few field tests on Navy ships and platforms were conducted.

Recently, the thrust of the magnetic particle inspection development has centered on the assembly of a kit that can be easily used and transported. During the investigation and field testing, it was found that MPI is usually used to investigate damaged or leaking areas of a ship's hull for cracks in hull welds or highly stressed areas such as rudder castings. The Navy's requirements for underwater MPI to respond to emergency situations is exemplified by the recent case of an old Navy destroyer that was found to have a crack in her bottom plating. The crack ends had been drilled and repaired from the inside, but the quality of the repair was questioned. NCSC performed an underwater MPI inspection on the cracked area and found that the arresting holes had not intersected the ends of the crack. As a result, the repair was judged unsatisfactory and the ship was taken out of service. This is a good example of the type of underwater MPI work for which the kit is designed.

The prototype underwater magnetic particle inspection kit (Fig. 1) includes the following components:

Two pairs of permanent magnets
Squeeze bottles of underwater magnetic particles in water solution
Underwater tape holder/writing tablet
White poly tape to pick up particles that show crack indications
Underwater battery-operated black light
Steel carrying case

To conduct an underwater magnetic particle inspection using this kit, the magnets are placed across the area to be inspected, and then particles from the squeeze bottle are applied. Excess particles are fanned away by hand, and the black light is used to illuminate the area for easy detection of concentrations of fluorescent particles indicating cracks or other magnetic discontinuities. If a crack is found, a piece of tape is taken from the tape holder/writing tablet and pressed onto the area to pick up the particles in their crack indicating pattern. Notes are made on the board about the location of the crack and any other important data. After picking up the crack indication or if no indication is found, the magnets are stepped along the area to be

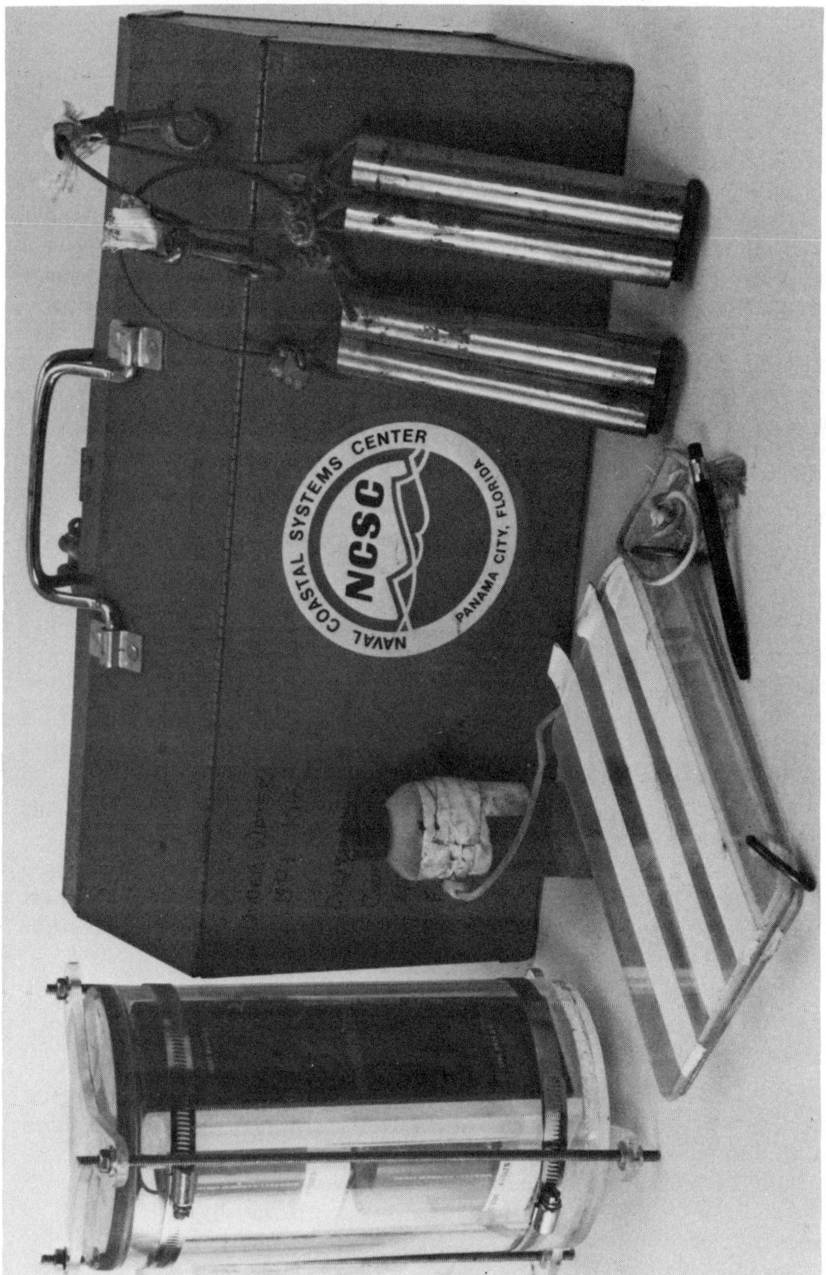

Fig. 1 – Prototype underwater magnetic particle inspection kit

inspected and the process is repeated. Two pairs of magnets are used so that one set of magnets is always attached while the others are moved, thus preventing the diver from losing his place along the weld or area being inspected. As this is being written, a field test of the kit on a Navy platform is underway and a preliminary set of procedures is being written.

Stereophotography

The underwater stereophotography system was developed to give the Navy an easy and inexpensive way to get quality photodocumentation of underwater structures. Most Navy divers have little or no experience with photography. The problem of attaining quality photography is further complicated by the poor underwater visibility found in most Navy harbors. The solution to these problems was found in using a readily available underwater camera and attaching it to a clear water box which, in addition to providing clear water between the camera and the surface being photographed, also provided a constant standoff distance. A strobe light was then mounted on the box to provide a constant light level for each photograph. The constant standoff distance and constant light level allow the camera to have focus, F-stop, and exposure time preset on the surface to predetermined settings. This allows the diver with no previous photographic experience to take the stereo system underwater, place it against the surface to be photographed, snap a picture, shift the camera to the second position, and snap the second exposure. When the diver surfaces, he has a roll of perfectly executed photographs. When the color photos are developed into 3-1/2 x 5-inch prints and mounted as stereo prints, an approximately full-size, three-dimensional picture of the photographed area is presented from which variations in elevation (depths of pits, etc.) can be measured for engineering analysis.

Figure 2 shows the stereophotographic system ready for use. This system has been used to document marine fouling and corrosion damage on naval ships as well as fixed structures such as bulkheads and offshore platforms. Recently the system was used to document the cleaning effectiveness of high-pressure water jets on propellers. The pictures allowed graphic proof of the cleaning that had been accomplished.

Ultrasonic Thickness Gauging

The ultrasonic thickness gauging system was developed to provide an underwater thin-hull survey capability that would be equivalent to or better than the surveys performed in dry dock. Thin-hull surveys are conducted on naval ships as they get older to determine steel thickness of critical hull plate areas, usually midships machinery spaces where hull stresses and corrosion are greatest. By performing thin-hull surveys with the ship afloat, well in advance of scheduled in dry dock maintenance periods, required plating renewal can be planned.

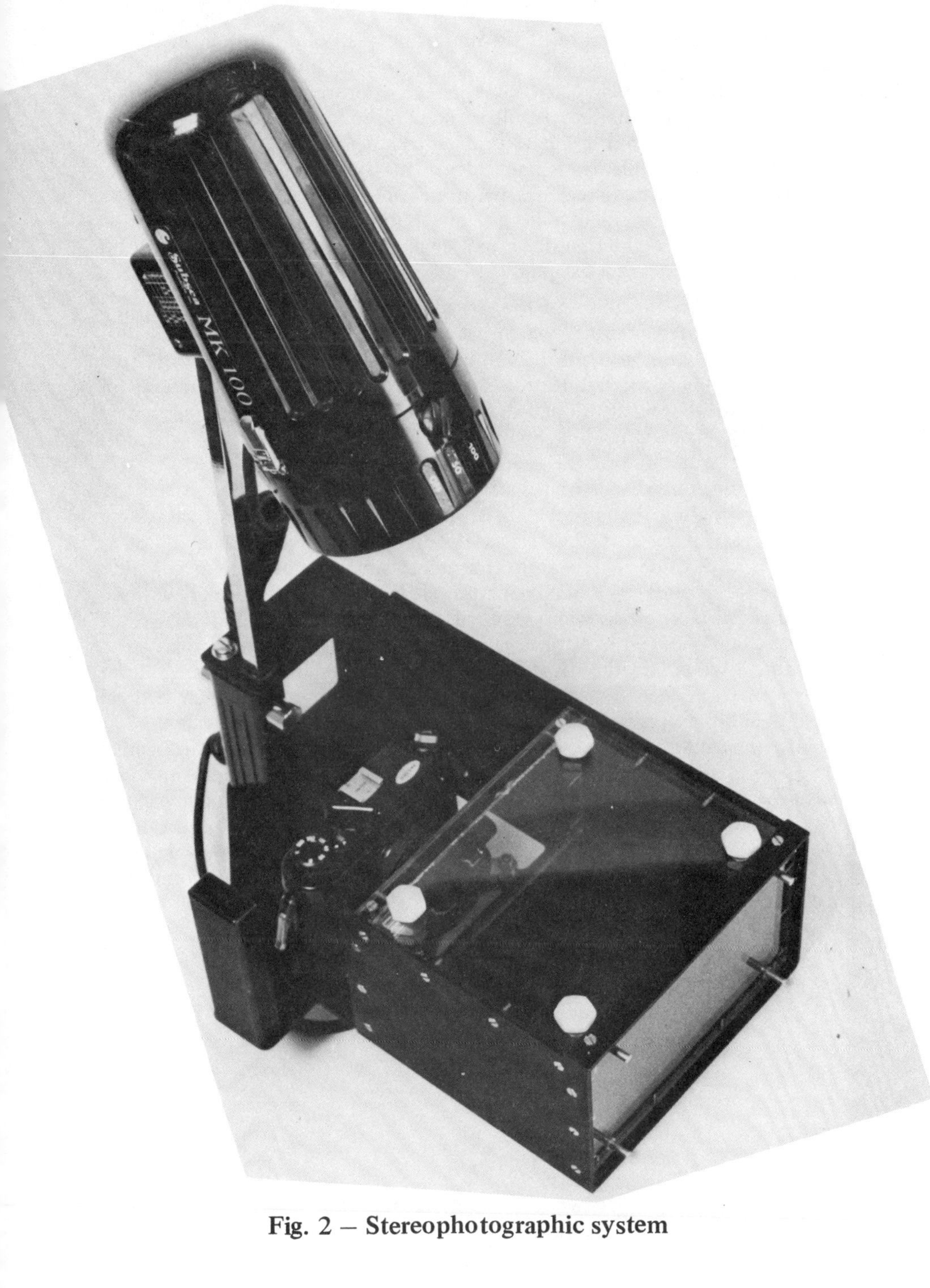

Fig. 2 – Stereophotographic system

Gaining the confidence of the nondiving decision makers in the thickness readings taken underwater was of paramount importance in gaining acceptance for the system. NCSC chose a team approach to thickness readings with a Navy diver positioning the ultrasonic transducer on the hull in response to commands from a topside ultrasonic technician who operates a fairly sophisticated ultrasonic testing machine. This system alone can duplicate the quality of thickness readings performed in dry dock. The addition of a computer to the system to take not one but 1,000 readings at a location while the diver rubs the transducer in a small circle on the hull is a significant improvement in thickness gauging.

The system as shown in Fig. 3 includes an immersion transducer in a special underwater housing, which feeds to a Nortec ultrasonic machine, which in turn feeds into a HP 9825 minicomputer through an A to D converter. The immersion transducer is more sensitive than the standard contact transducer. The specially designed NCSC housing is shaped, approximately beer can size, for easy handling by divers. The housing also provides a standoff distance for protection of the transducer and to get the signal used in thickness measuring outside the transducer's near field. By using the more sensitive immersion transducer and providing it with a standoff, a very sensitive system results. This means that the surface condition of the steel plate is not as critical as with other underwater ultrasonic systems. Paint and corrosion do not have to be removed. Usually surveys are done after hull cleaning with no additional surface preparation required. The ultrasonic machine, a Nortec 131D, is equipped with both a cathode ray tube and digital display. The operator, after telling the diver where to take the reading, makes whatever adjustments are needed as the diver slowly moves the transducer in a small circle. When the operator has the signal he wants, he activates the computer to take the 1000 readings directly from the ultrasonic machine. Meanwhile, the operator has entered the location, original plate thickness, and any other pertinent data into the computer. The computer stores the data and each of the 1000 readings on magnetic tape for later analysis. At the completion of the survey, a printer and plotter are joined to the computer, and whole pages as shown in Fig. 4 are automatically printed in standard report format. In addition, a plot of each of the 1000 thickness readings can be plotted as shown in Fig. 5 if desired. This particular plot shows a fairly constant thickness in the area and only a small number of uncoupled readings shown as 1.00 inch.

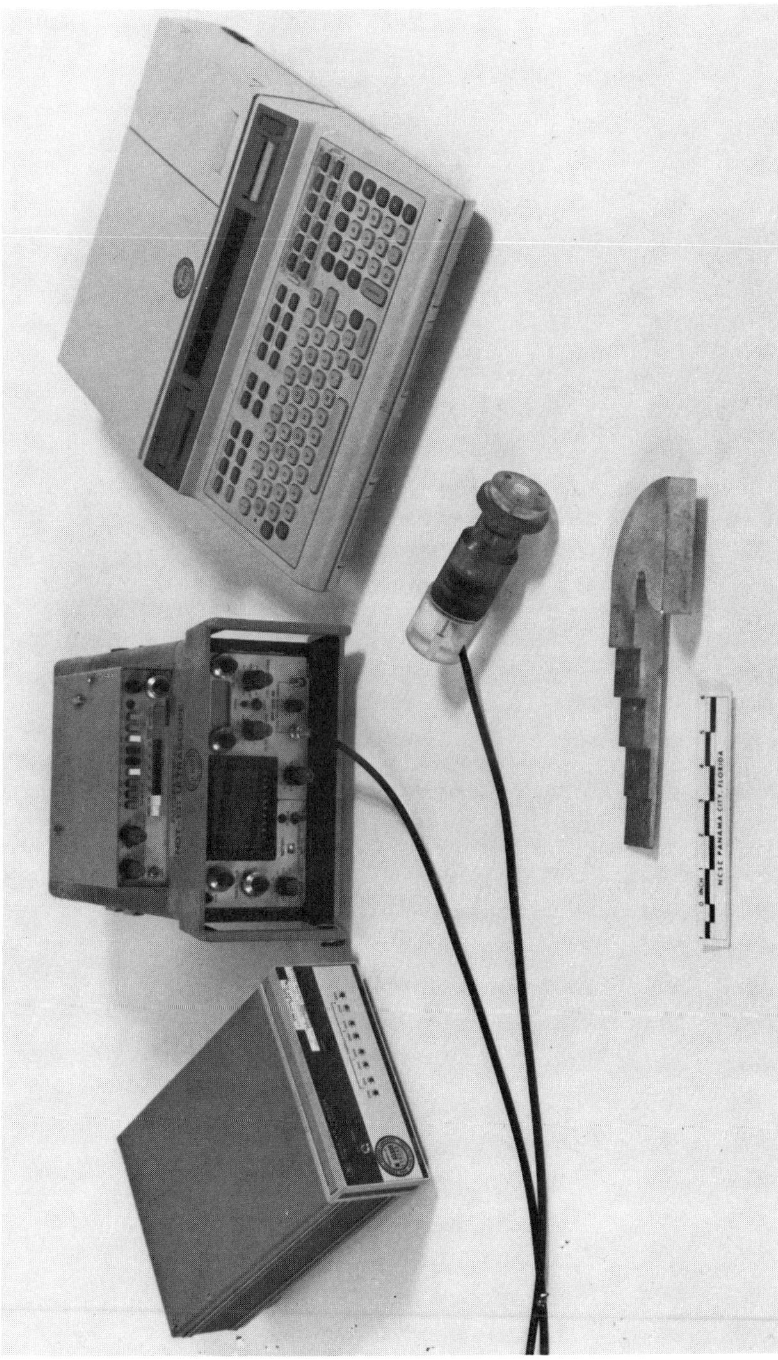

Fig. 3 – Ultrasonic thickness gauging system

USS BRYCE CANYON AD-36 7/80

BETWEEN FRAMES 112 AND 113

LOCATION	DESIGN THICKNESS	MEAN	MAX	MIN	PERCENT THINNED
frame 112+15", W/L(50')	0.625	0.615	0.628	0.602	2
frame 112+15", FG+12 port	0.625	0.610	0.634	0.590	2
frame 112+15", FG-12 port	0.625	0.608	0.626	0.582	3
frame 112+15", EF+12 port	0.625	0.546	0.566	0.476	13
frame 112+15", EF-12 port	0.688	0.675	0.688	0.650	2
frame 112+15", ED+12 port	0.688	0.671	0.690	0.664	2
frame 112+15", ED-12 port	0.688	0.674	0.686	0.636	2
frame 112+15", CD+12 port	0.688	0.682	0.701	0.660	1
frame 112+15", CD-12 port	0.688	0.667	0.678	0.654	3
frame 112+15", BC+12 port	0.688	0.663	0.674	0.638	4
frame 112+15", BC-12 port	0.688	0.679	0.695	0.620	1
frame 112+15", AB+12 port	0.688	0.678	0.690	0.656	1
frame 112+15", BA-12 port	0.688	0.639	0.656	0.620	7
frame 112+15", A/FK+12 port	0.688	0.656	0.670	0.636	5
frame 112+15", A/FK-3 port	0.875	0.837	0.853	0.791	4
frame 112+15", KEEL CL	0.875	0.862	0.889	0.859	1
frame 112+15", A/FK-3 stbd	0.875	0.851	0.861	0.821	3
frame 112+15", A/FK+12 stbd	0.688	0.672	0.705	0.638	2
frame 112+15", AB-12 stbd	0.688	0.666	0.686	0.644	3
frame 112+15", AB+12 stbd	0.688	0.684	0.694	0.652	1
frame 112+15", BC-12 stbd	0.688	0.675	0.690	0.636	2
frame 112+15", BC+12 stbd	0.688	0.680	0.737	0.660	1
frame 112+15", CD-12 stbd	0.688	0.678	0.697	0.620	1
frame 112+15", CD+12 stbd	0.688	0.670	0.699	0.622	3
frame 112+15", DE-12 stbd	0.688	0.679	0.692	0.670	1
frame 112+15", DE+12 stbd	0.688	0.677	0.688	0.660	2
frame 112+15", EF-12 stbd	0.688	0.659	0.676	0.652	4
frame 112+15", EF+12 stbd	0.625	0.635	0.644	0.598	-2
frame 112+15", EF+12 stbd	0.625	0.572	0.596	0.518	8
frame 112+15", EF+12 stbd	0.625	0.138	0.194	0.099	78
frame 112+15", EF+12 stbd	0.625	0.118	0.147	0.099	81
frame 112+15", FG-12 stbd	0.625	0.630	0.638	0.608	-1
frame 112+15", FG-12 stbd	0.625	0.625	0.642	0.606	0
frame 112+27", FG-12 stbd	0.625	0.618	0.632	0.610	1
frame 112+15", FG-12 stbd	0.625	0.613	0.634	0.592	2
frame 111+27", FG-12 stbd	0.625	0.614	0.630	0.564	2
frame 112+15", FG-12 stbd	0.625	0.500	0.506	0.488	20

Fig. 4 – Computer-generated ultrasonic thickness gauging report page

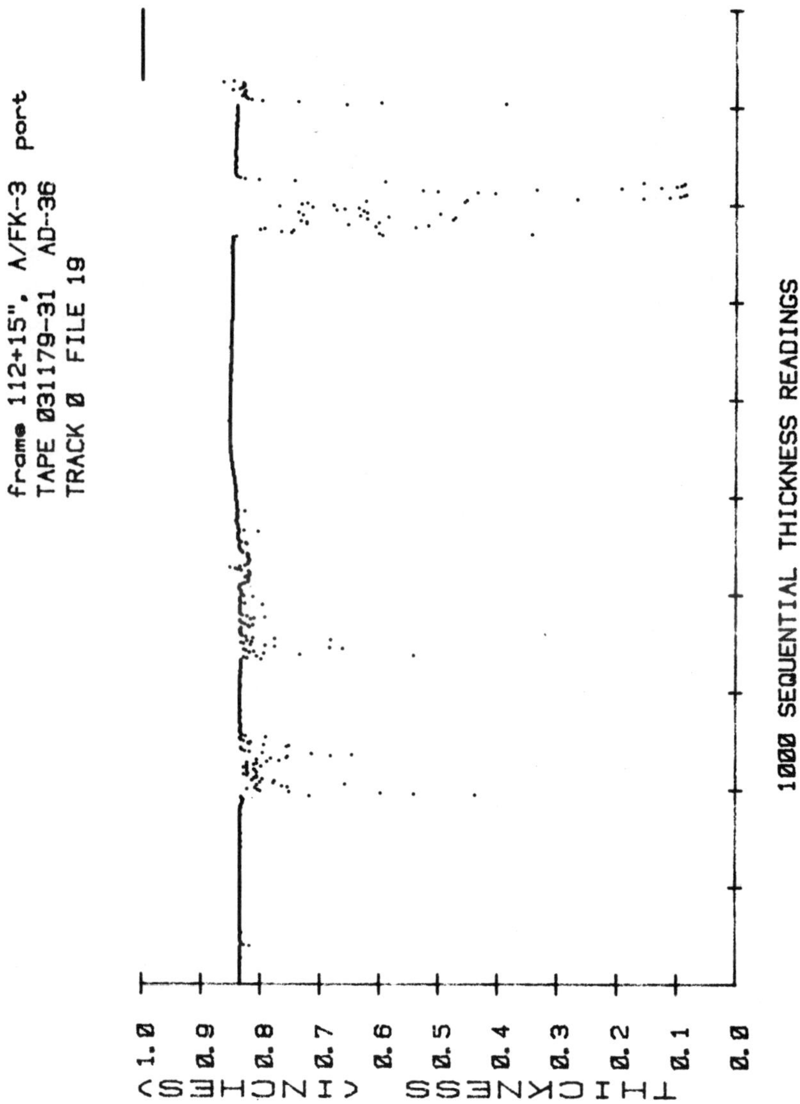

Fig. 5 — Computer-generated plot of 1000 thickness readings taken at one location

The system has been field tested on approximately 15 naval ships. At present the procedures manual has been prepared, and the system is undergoing final approval for Fleet use. It is anticipated that the system will be well accepted in the Navy due to these attractive features:

(1) The involvement of qualified NDT personnel who can interpret and remark on the screen presentation

(2) The presence of a computer (psychologically this seems to make results more credible)

(3) The engineering usefulness of the data that comes from taking a large number of samples at each location

(4) The very minor effect the diver has on the data

(5) The use of immersion transducers, which perform much better than contact transducers on corroded steel

(6) The minimal requirements for preparation (no deballasting or pumping tanks, etc.) and freedom of access to all parts of the underwater hull

(7) The timeliness of final reports, which are computer generated, and are usually sufficient to specify hull plate repair requirements in detail

(8) The minimal interference of underwater surveys with operating schedules (often these surveys are scheduled around waterborne hull cleanings)

Ultrasonic and Electromagnetic Flaw Detection

Ultrasonic and electromagnetic flaw detection are being studied at NCSC for use in the underwater ship's husbandry program. Flaw detection is well developed for surface use, but, because it requires a high degree of operator skill, it is not readily adapted to underwater use. Accurate positioning and orientation of the transducer is required as well as concentration by the operator, both of which are difficult for the diver in the hostile underwater environment. NCSC is pursuing both ultrasonic and hull effect flaw detection methods at present because each has an electrical output that can be computer processed on the surface under the supervision of a qualified NDE technician.

A crude experimental model of the hull effect device has been assembled in the laboratory as shown in Fig. 6. The system has shown promise and will undergo further development during the next year.

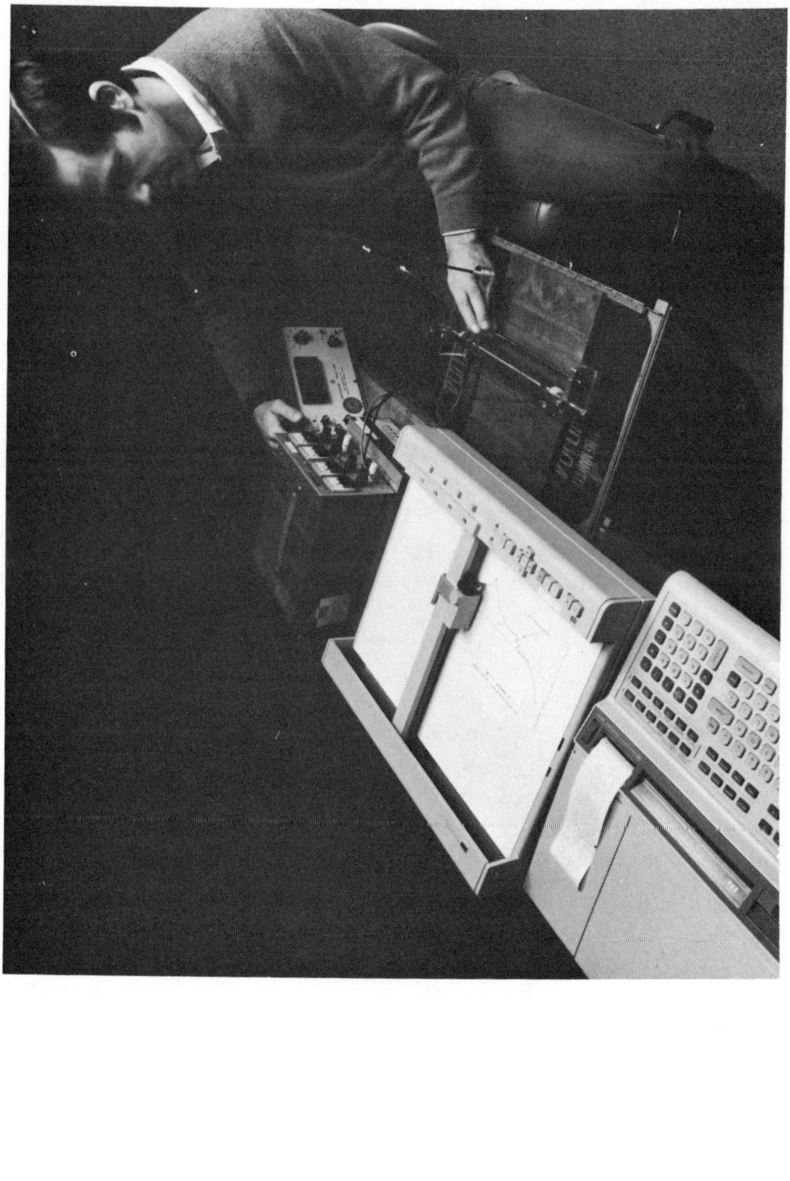

Fig. 6 — Laboratory setup for hull effect flaw detection

Figure 7 shows a conceptual drawing for a flaw detection and evaluation system. The concept is for the diver to attach the three-leg positioning device to the hull at a specific location, and then to move the transducer randomly over the area of interest. Topside, the computer receives position data from the positioning arm (each joint in the arm has a form of electrical rheostat that indicates the angular displacement for the joint). The computer takes this information plus the known length of arms and calculates the transducer position and orientation. Then the computer interrogates the ultrasonic or hull effect instrument for its flaw indication. The results of position and flaw can then be plotted by the computer giving a graphic representation of the flaw.

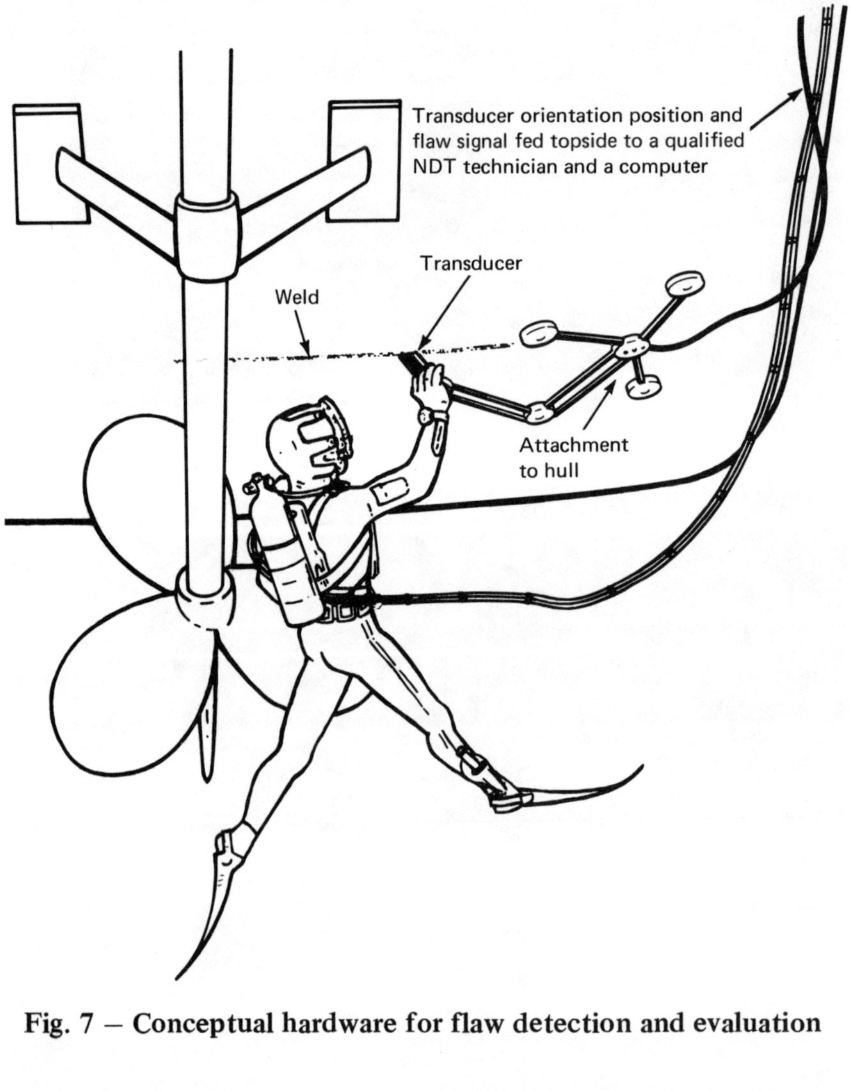

Fig. 7 – Conceptual hardware for flaw detection and evaluation

Summary

NCSC has developed workable underwater stereophotography, magnetic particle inspection, and ultrasonic thickness gauging systems that have been proved effective in field experiments. The stereophotography system development has been completed and is available for Fleet use. The ultrasonic thickness gauging system has been used on many thin-hull surveys of naval ships and presently is in the final stages of approval prior to its general use by the Fleet. The magnetic particle inspection system is evolving into an easily transported and easily used kit. A prototype kit has been assembled and proved effective at the laboratory. Ultrasonic and electromagnetic flaw detection is under development and will be a prime area of investigation during the next year.

A Progress Report on Underwater Welding and Welder/Diver Training by the U.S. Navy

Thomas J. Dawson and Dale G. Uhler
Naval Facilities Engineering Command

In an article in the NAVSEA Journal, June 1979, entitled "Underwater Welding—A Diver's Operational Dynamic Tool," the Navy's status was evaluated. It was pointed out that while the Navy was behind commercial firms in state-of-the-art underwater welding, considerable progress was being made to catch up. Giant steps have been taken not only to close this gap, but to forge ahead and re-establish the Navy in its rightful place as a leader in underwater welding and diving.

During the past year the Navy has:

(1) Completed the evaluation of a new stainless steel shielded metal arc (SMA) electrode developed by the International Nickel Company. This electrode can be used in underwater welding, when waterproofed, and in the field welding of difficult-to-weld structural alloys in air (terrestrial welding).

(2) Evaluated at the Civil Engineering Laboratory (CEL), Port Hueneme, California, SMA terrestrial welding electrode waterproofing materials and techniques.

(3) Obtained the patent rights to Olefin shrink tubing, a convenient, reliable waterproofing material. Considering that commercially available electrodes for underwater welding are limited, the application of Olefin shrink tubing to commonly used, everyday electrodes makes it possible to have an unlimited variety of electrodes for underwater use, assuming the testing program discussed in (4) below is successful. Another application of the tubing, for better efficiency, is the waterproofing of oxygen (air) cutting and gouging electrodes. Under field conditions and when compared to the other waterproofing materials investigated, the tubing has a long, reliable shelf-life. Also, it can easily be applied in a few minutes using the electrode drying ovens and a few other tools (job-site available). Little training in its use is required.

(4) Initiated a program to conduct tests by the Welding Engineering Laboratory at the Norfolk Naval Shipyard to determine moisture absorption rates. Electrodes that have been waterproofed with Olefin type shrink tubing will be subjected to water pressures to 200 feet for various time periods (Fig. 1). They will then be used to make welds. The test welds will be analyzed for mechanical properties and for any metallurgical changes resulting from the waterproofing of electrodes. All test welds will be subjected to nondestructive evaluation (NDE). This work will be completed this year.

(5) Added MAPP gas to the welding and cutting manual.

(6) Completed the revision of the Navy Underwater Cutting and Welding Manual (NAVSEA 0929-LP-000-8010).

(7) Developed, through the Welding Engineering Department, Welding School and Rigging Department (Diving Group) of the Norfolk Naval Shipyard, the training facilities for a proposed comprehensive underwater welder training program.

The existing terrestrial Welding School, the experienced welder/divers of the Rigging Department, the Welding Engineering Department, NAVFAC NAVSEA Ocean Engineering Directorate, SURFLANT, SURFPAC, and Chief of Navy Technical Training (CNTT) are cooperating to develop a functional underwater wet welding school. This will be the first school of its kind in the world and is probably the most significant achievement the Navy will make to upgrade its underwater welding capability. Through academic study and laboratory training, Navy Diving School trained divers with welding aptitudes will achieve competency in underwater welding. The aim is a thorough program of increasing encumbrances and utilization of the diver's skills so that he will be able to produce quality welds underwater at various depths and within the limitations of his support equipment.

The candidates for training will be designated from graduates of the Naval School of Diving and Salvage (NSDS). It is hoped that these divers will be those with extra enlistment tenure due to the time required for the multiple skill training. The attractiveness of the pay and the shortage of trained welder/divers presents the same problem of sailor retention that existed with nuclear welders a decade ago.

Upon arrival at the Welding School, the designated candidate will undergo tests and evaluation to establish competency as a terrestrial welder. The trainee will immediately be started on a program to strengthen weak areas of skills required for underwater welder/diver training. Some academic training may collaterally begin at this time.

Fig. 1 – Test tank assembly for pressure testing Olefin shrink tubing and other protective coatings for effectiveness at varying pressures and times

The underwater welding will be started in an underwater simulator (Fig. 2). This simulator is equipped with ring-stiffened sleeves of a diving suit and a large viewing window so that the instructor and trainee can have a clear view of the welding arc, and fusion welds in three positions—downhand or flat, vertical, and horizontal—can be made in the simulator. The pressure in the simulator is limited to the static pressure of the water head. Low-pressure hyperbaric welds could also be made in this simulator under inerted conditions. Note the grab bar above the viewing window for use by the trainee for stability while viewing the weld being deposited.

Figure 3 shows the interior of the simulator at front, with the sleeves and hand gloves withdrawn.

Figure 4 shows the interior with details of the practice or training plate racks. The platform at the bottom front center is to house the welding electrode holder, scaling hammer, wire brush, and other needed hand tools. Note the light above the door for illumination of interpass weld quality and weld completion by visual inspection.

Figure 5 shows the welding current off-on control and electrical wire to feed the inspection light.

Figure 6 is the trainee's view of a weld being deposited in the vertical position. The arc has just been restarted, hence the small volume of bubbles rising from the arc area. The relative size of the arc-created bubble is clearly defined. The diver is thus comfortably seated with the minimum encumbrance (diver suit sleeves with gloves) necessary for the start of underwater welding training. The welder trainee will be trained here to make welds in the three positions outlined previously.

Six of these simulators are being constructed so that six trainees will be able to weld at the same time. Continuous welding is impractical. Thus, two groups are planned for simultaneous training. One group of six will be welding while the other six are attending academic training.

Satisfactory achievement of skills to complete quality welds in the flat, horizontal, and vertical positions would move the training to the ten-foot water tank. Here the trainee would learn to weld overhead. He would be further encumbered by a complete diving suit, with simulated increased depth, low temperatures, flow rates at different levels, and turbid conditions.

Figure 7 shows the existing ten-foot tank. A platform is planned for installation at tank-top level. Lockers will be placed along the two walls shown with provisions for a suit-up area and diving support equipment.

The large view ports on two sides of the tank allow excellent observation of progress in the tank. Note the brackets for external lights outside the tank for viewing interior progress.

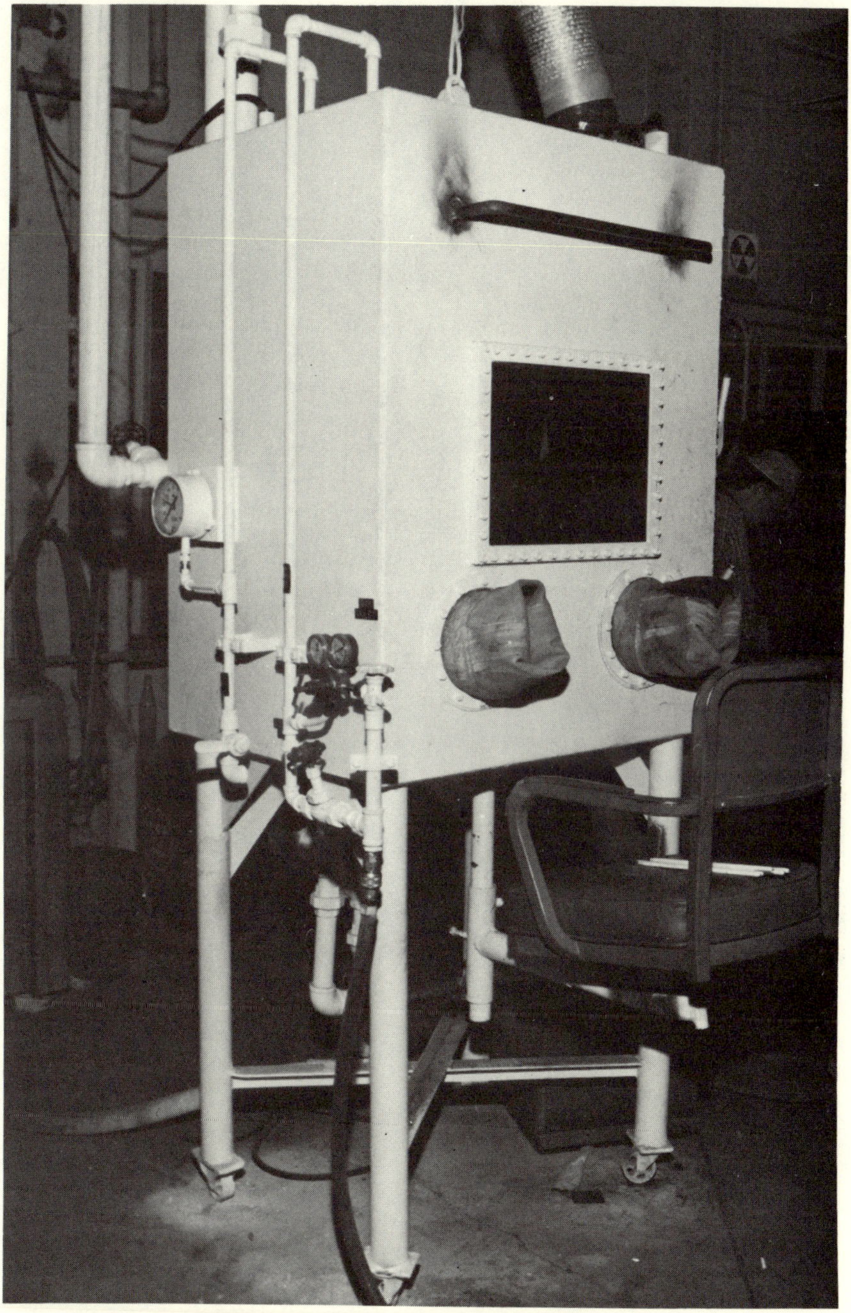

Fig. 2 – Underwater simulator

Fig. 3 – Inside of simulator showing gloved sleeves withdrawn, window, and worklight

Fig. 4 – Interior details of simulator, showing test plate, racks, etc.

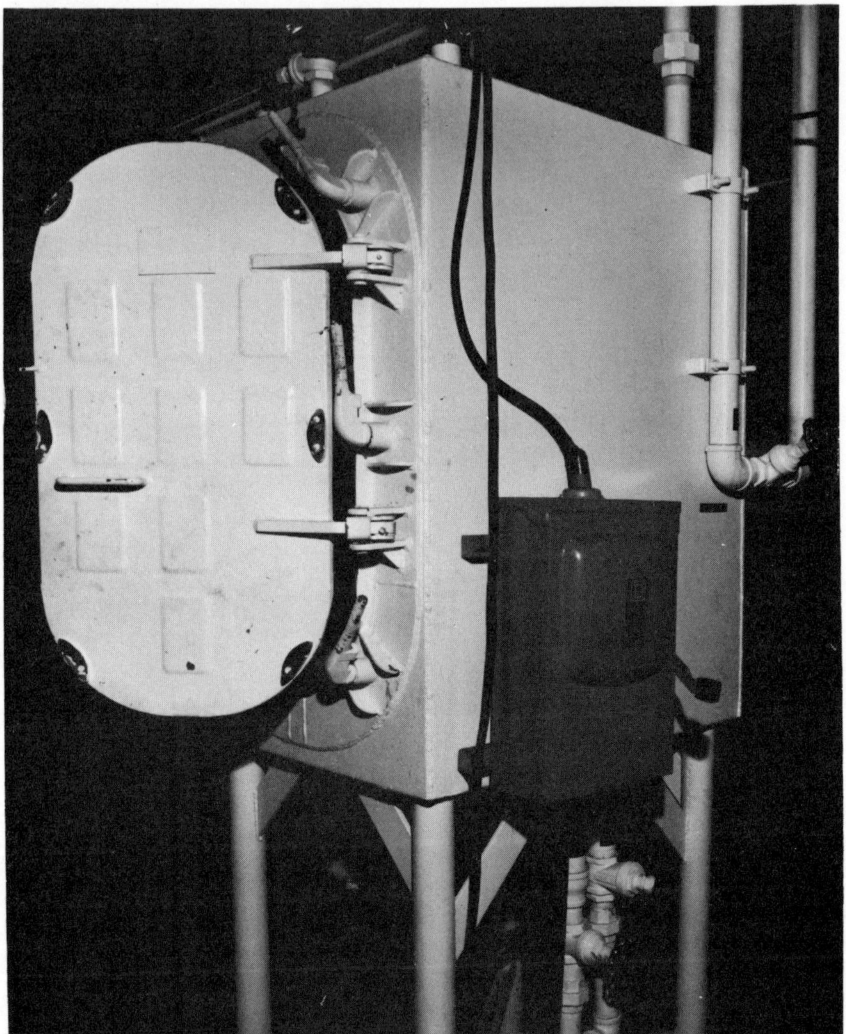

Fig. 5 – Welding current on-off control plus valving and piping for rapid flooding and emptying

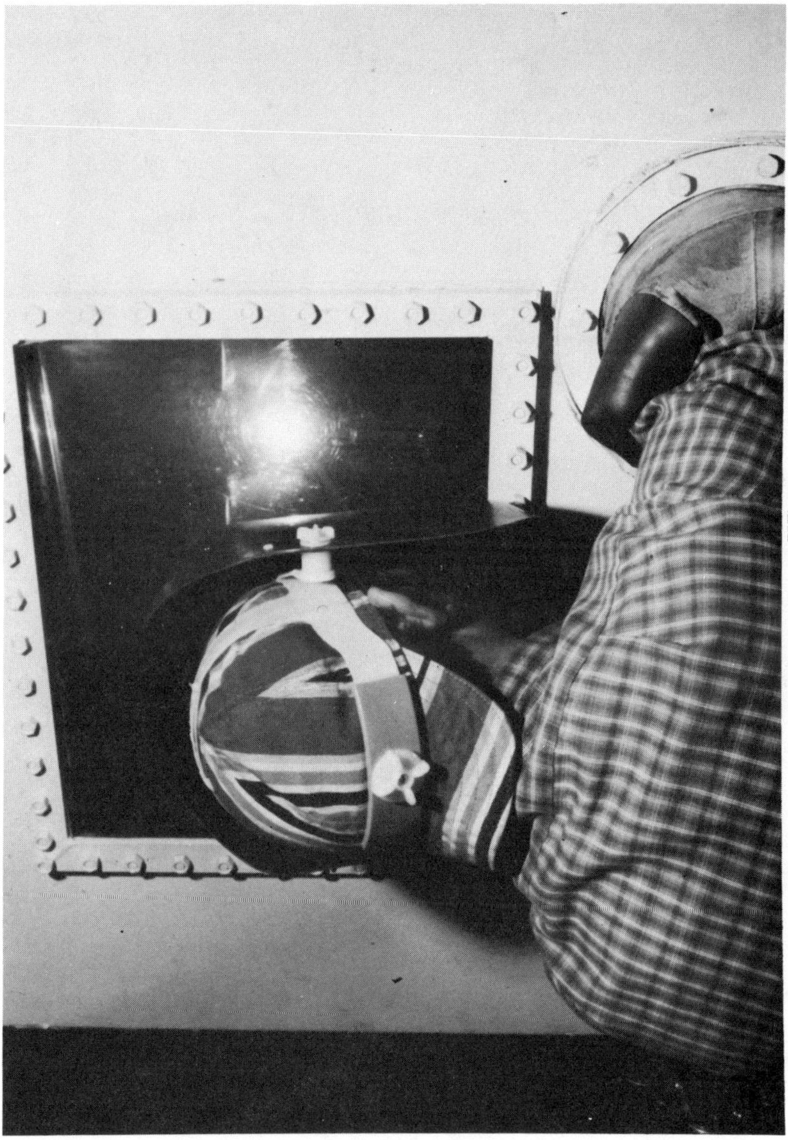

Fig. 6 – Trainee's view of weld being deposited in vertical position

Fig. 7 — Existing 10-foot tank

A tank cover is available that will allow pressurizing to simulated depths greater than ten feet. Here, the diver will be trained at elevated pressure so he will be cognizant of the arc characteristic changes that occur with pressure. In addition to varying pressures, the trainee will also weld in the tank at varying water temperatures and water velocities simulating job-site conditions. The water in the tank is also capable of being made turbid if needed for training, test work, research, or special procedure development.

When the skill for flat, vertical, horizontal, and overhead welding has been developed to the satisfaction of the instructors, the welder/diver will be considered ready for qualification testing. The qualification test welds will be made from a 30 foot depth platform in an *in situ* environment where river currents will be simulated.

Special evaluation of underwater welding techniques and materials can now be performed in facilities not available at other naval activities.

The divers will be trained using electrodes and welding variables developed by the Welding Engineering Department of Norfolk Naval Shipyard, NAVFAC, NAVSEA, SURFLANT, and the Chief of Navy Technical Training. Procedures will be given the welder/diver trainees for steel with carbon equivalent of 40 and less as well as structural materials with a carbon equivalent greater than 40. They will also be instructed in the art of waterproofing SMA electrodes and will be trained using approved Navy diving support equipment.

The present schedule calls for initiation of a pilot training program during FY 81. The number of welder/divers to be trained each year will be based on an assessment of the Navy's needs for such skilled personnel.

An active research and development program is scheduled for FY 80-81. The Civil Engineering Laboratory will be developing and evaluating a diver back pack MAPP gas-oxygen emergency unit or kit. The quality of welds made with fluorocarbon self-shielding flux cored welding filler material will also be evaluated. The follow-on program in FY 81 includes the evaluation of using the above welding filler material with semiautomatic welding equipment in the wet or in a minihabitat with the welder in the wet.

The National Research Council, Structural Division of the Ships Structural Committee, is investigating the fatigue of underwater welds containing various discontinuities with special emphasis on tubular structures.

The Civil Engineering Laboratory at Port Hueneme has an extensive program on underwater NDE procedures and equipment. The same is true for the Naval Coastal Systems Center at Panama City, Florida.

Thus, through continued research, evaluation, and training, the Navy hopes to again become the front runner in the field of underwater welding.

Description and Discussion of the State of the Art in Underwater Nondestructive Testing

Bruce J. Sylvester
Sylvester Underseas Inspection

Underwater nondestructive inspection and testing is receiving a great amount of interest. Primary attention in the United States is being focused by United States Navy on hull and wharf maintenance and by the offshore oil industry on the inspection of drilling rigs, fixed platforms, and underwater pipelines. All of these structures are subjected to the harsh environment of salt water, constant cyclic loading, and intermittent high service stress. Underwater nondestructive testing as a method to determine initial material breakdown well in advance of structural failure is being investigated, evaluated, and utilized with varying degrees of success.

Nondestructive testing (NDT) as described in this paper will mainly concern itself with the common methods of underwater steel inspection with the emphasis on magnetic particle inspection (MT), radiography (RT), and ultrasonic inspection (UT). The paper will describe the strengths and weaknesses of MT, RT, and UT when used underwater by divers with a description of the various underwater influences on each. Other types of nondestructive testing briefly described are underwater visual inspection, photography, corrosion potential measurement, acoustic emission, liquid penetrant inspection, and the use of remote controlled underwater vehicles.

Acoustic Emission

Acoustic emission is a nondestructive monitoring method. Several piezoelectric transducers are affixed to the structure being monitored, in this case an offshore platform or perhaps an underwater pipeline. Normal flexing of the structure caused by wind loads or wave action causes the structure to "creak." Any discontinuities within the structure will emit characteristic sounds detectable by the piezoelectric transducers. This information is run through topside electronic processing, which provides the operator information to interpret the approximate location of the sound origin. If a defect opens up, such as a fresh crack or when an existing defect propagates, a new or different sound emission will be detected by this system. To pinpoint and characterize the flaw requires additional investigation by one or more of the other methods described in this paper.

Acoustic emission is still in the development stages for use underwater and, to the author's knowledge, is not yet used routinely for any offshore structure other than in experimental situations.

Corrosion Potential Measurement

Almost all underwater structures are protected by one variety or another of a corrosion protection system. This may be through impressed current, sacrificial anodes, or a combination of both. In all cases, however, the voltage generated must be adequate to protect the structure. The amount of surface area to be protected, geometric configuration of the structure, the water temperature, oxygen content, salinity, and water velocity are important factors that influence the proper protection of an underwater structure.

The electric potential may be measured around the structure through the use of standard half cells modified for use underwater. Often referred to as a "CP meter," they are generally handled by divers who will bring the half cell to various locations on the structure to measure the voltage potential through the sea water. Several varieties of equipment for this purpose are currently available. Some units have a silver/silver chloride or copper/copper sulfate reference electrode connected by a long wire that attaches to the structure either on the surface or near the location where the measurement is being taken. The half cell is held near the structure by the diver, completing the circuit, and the electrical potential through the seawater is measured by a high-input impedance voltmeter. One unit is shaped like a gun and the diver simply presses the nose of the gun against the structure to complete the circuit. A silver/silver chloride half cell is located within the framework of the gun close to the contact point, and a large LED readout on the back of the gun reads out the voltage measurement to the diver. Although this unit is more portable underwater, the paint or coating system must be disturbed each time a measurement is taken, a small penalty if the structure is properly protected.

Another method is to keep the electronics on the surface with the test connection wire attached to the structure and another long wire attached to the reference cell. A diver or RCV may be used to bring the reference electrode to the measurement location and the voltage potential is observed and recorded on the surface.

Although satisfactory results can usually be obtained through marine growth, best results are obtained on cleaned surfaces.

Underwater Photography

Underwater photography is also a well known method of underwater nondestructive inspection. The difficulties with this test lie in operator skill in handling the equipment and photography in turbid water conditions.

Operator skill may be compensated for by new cameras that have a large depth-of-field so that the focus adjustment is less critical. A self-regulating strobe flash will automatically adjust for a properly lit exposure. These cameras only require a diver or RCV operator to "aim and shoot." Often a probe or standoff will protrude from the front of the camera to help the diver accurately maintain the focus distance with the correct exposure setting.

Photography in turbid water conditions is accomplished by placing a layer of fresh water in a chamber between the camera and object being photographed. Much work is being performed by the U.S. Navy in Panama City, Florida, utilizing this method.

Visual Inspection

Visual inspection by divers is well known and self-explanatory. It is the most common of all nondestructive tests. Topside observation and permanent recording of a visual inspection can be provided by underwater television capability.

Liquid Penetrant Inspection

This common land-based nondestructive test is practically nonexistent in underwater use and rarely used even in dry habitat situations. Low temperatures, humidity, and generally unsuitable material surface conditions render the test highly impractical in the underwater inspection situation.

Magnetic Particle Inspection

Magnetic particle inspection, as used on land, is a fast, revealing test for the detection of surface discontinuities and flaws such as cracks, surface laminations, and similar defects. The test will work even on tightly closed defects including defects that are not visible to the unaided eye. Some near-surface (1/4 in.) flaws can often be detected by this method. The test is generally administered by a technician and a helper; however, it may be performed by one man given semi-automated MT equipment. The basic principle of the test is to induce a magnetic field through the material under evaluation. A discontinuity will cause a flux leakage within that field and will attract ferritic particles to the flux location. These particles will then form over the discontinuities such that they may be seen and recognized by the inspecting technician. The magnetic field may be induced by using two electromagnets or permanent magnets, or by inducing a high-amperage low-voltage current through the material with the resultant magnetic effects.

This test is successfully used underwater in some situations. Typically, a permanent magnet or low-strength electromagnet equipment will be used underwater. Use of this type of equipment is due to its smaller size and greater mobility; however, the lower field strength reduces the effectiveness of the test. In the North Sea, operators use more conventional watertight rectified units that often weigh more than a ton. These are lowered by a crane to the inspection location for use by the divers. The inherent difficulties in the use of this equipment are bulk and immobility. This type of power pack, however, does allow a much greater field strength and therefore a more sensitive test for defects.

Various factors must be considered when planning the use of underwater MT. If an electric current is to be induced into the material, the surfaces must be clean enough of marine growth to allow electrical contact. Second, more than one diver is generally required to manipulate all the items needed to perform the inspection. A pair of prods generally held together on a common yoke must be held against the structure for the duration of the test. Often, the diver is "suspended" in the water with nothing to brace against to allow proper prod pressure against the material. Therefore, some type of brace or restraint must be provided for the diver holding the prods against the structure. The magnetic particles must be introduced into the inspection area so that they may be attracted to potential defects. This can generally be accomplished by premixing in a water or detergent solution ferritic particles designed for this use. The solution is then squirted out of a bulb into the area where the test is being performed. More often than not these particles are fluorescent by design so that a black light may be used to more rapidly identify defect formations of the particles. Therefore, a second diver will be required to hold a black light.

In a poor underwater visibility situation, the test certainly has additional difficulties or may be rendered unusable. Also, fast moving currents make the test difficult to administer both from the standpoint of handling bulky equipment, maintaining position, and also in keeping the magnetic particles in the local area under examination.

A final consideration is that there is no good method for topside personnel to monitor the application of the test. The client or interested party must rely entirely on the skill and discipline of the divers to perform the test satisfactorily without supervision or direct observation.

In spite of the above difficulties, there is much enthusiasm for use of this test underwater in the North Sea. Relatively little underwater MT has been performed in United States waters. The United States Navy has found it to be a good test under laboratory and in pool conditions. A recent evaluation, however, by a major oil company in Scotland under average underwater working conditions, i.e., slight current, medium visibility, and cold water, disclosed MT to be most questionable in its ability to locate known defects, as was reported to the author by representatives involved in the test.

Magnetic particle inspection performed in an underwater dry habitat may be considered essentially as good a test as when performed on land. The primary problem here is bringing the MT equipment to the habitat, which may be located as deep as 200 or 300 feet underneath the surface of the water. The author's firm has performed inspection using electromagnetic equipment with acceptable results. The only other difficulty encountered is that of the high humidity environment of a typical underwater habitat. This causes most surfaces to be moist, so good distribution and free movement of the ferritic particles is a more complicated problem than on a dry surface. Again, mixing the particles in a slurry of water or detergent generally solves the problem.

Underwater Radiography

Radiography on land is one of the most commonly used and accepted tests in the fabrication industry. One of the primary benefits of radiography is the establishment of a permanent record, which allows later review by persons other than those persons who conducted the tests. From these radiographs two things may be established:

(1) The conditions of the materials under examination.

(2) An assessment of the sensitivity or quality of the test itself, allowing proper evaluation of the defect level observed, i.e. a poor radiograph will not show small defects, while a very sensitive sharp radiograph may be expected to show more detail.

This test is used underwater only in limited applications and does not enjoy the overall use that it does on land. The primary difficulties in using radiography underwater are:

(1) The deleterious effects of water, which attenuates and scatters radiation such that just several inches of water will diminish and unfocus a radiation beam so that reliable results are generally not obtained.

(2) The hazards in using radioactive devices are magnified in the underwater situation.

As a result of our studies at Sylvester Underseas Inspection, we have established that 9 in. of water path between a .1 in. x .1 in. source of Iridium 192 will scatter the radiation to the extent that a 2-2T hole is not perceptible through a 1/2 in. plate using Kodak M fine-grain film. This unfocusing problem has been overcome by some operators by installing a water barrier in the path of the radiation. This is generally accomplished by fabricating an air capsule, foam capsule, or similar water-excluding device that allows freedom from attentuation and scattering of the radiation beam between the gamma ray source and the specimen under observation. This particular technique offers some handling difficulties but in general works satisfactorily.

Radiography is not only used underwater to find weld flaws and defects but is also used actively in the detection and characterization of internal corrosion. For instance, internal corrosion in a tubular structure such as a wharf piling may be observed by taking a radiograph, double wall, of the wharf piling. Corrosion pits or corroded areas of the pipe show up as contrasting areas on the developed radiograph. Generally speaking, a lower level of sharpness is required for the observation of this type of material condition than is required to see weld defects, so a certain amount of diffusion of the radiation beam can be tolerated. The drawbacks of this test are the expense of the test and the relatively small area sampled. Generally speaking, however, several random samples of a dock or pipeline will often be quite revealing for the general condition of the entire structure. Some work has been performed in correlating the density differences in developed radiographs with actual material thickness. This method has met with mixed success from an accuracy point of view; however, from a practical point of view is quite adequate. To put it another way, it may not be possible to judge pit depths to within 10 or 20 mils; nevertheless, if the pitting is significant it is immediately apparent upon review of the radiograph.

Structural welds in tubular materials present a difficult problem. Unfortunately, most underwater structural materials are tubular in nature whether it be pipes to carry the product, the tubular structure of the platform or drilling rig, or the tubular structure of many piles. (Ship hulls are the prominent exception.) In many cases these tubular materials are filled with water or concrete. If the diameter is greater than 9 in. and the pipe is full of water or concrete, it is almost impossible to take a "code quality" radiograph of a weld or pipe wall deterioration.

An unexpected and happy exception to the above situation is a recent observation by this firm that an oil-filled 12-in. pipe with half-inch wall can be radiographed with excellent sensitivity. We attribute this success to the lower density of oil as opposed to water. Further work is under way to establish the radiographic limits of oil-filled pipes.

Another problem area for underwater radiography is in the extensive use of fillet welds underwater as opposed to butt joint welds in most structural situations. Almost all underwater repairs to structures will utilize fillet welds for simplicity and ease of installation. Unfortunately, fillet welds are generally difficult to examine by radiography due to their geometric configuration. Therefore, radiography enjoys much less success under the water than it does on the surface as a standard method of inspection.

The radiography of ships' hulls while in the water works well by use of the water barrier method or by keeping the radiation source in close enough proximity to the material under examination so that a great amount of diffusion is not allowed to occur. Naturally, the source must not be kept too close or the picture will be unfocused by "geometric unsharpness." A study by our organization has indicated that a gamma source, Iridium 192 .1 in. x .1 in., will give acceptable 2-2T radiographs using Kodak M fine-grain film through a 1/2 in. plate from a distance of no closer than 5 in. to no further than 9 in. from the material under examination.

Another consideration is that radiography requires access to both sides of the material, a situation not always possible in many underwater situations.

Radiography performed in dry underwater habitats is a routinely performed effective test for the examination of welded tubular butt joints. Usually in the underwater butt joint weld of a pipeline, the water or product has already been evacuated to allow welding of the joint.

Underwater Ultrasonic Inspection

Underwater ultrasonic inspection to detect corrosion via thickness measurements and to interrogate materials and welds for flaws is finding increasing use underwater. Ultrasonic testing shares some of the same handicaps underwater as it does on land, that is, a high level of knowledge to interpret the ultrasonic indications and general distrust by the fabrication industry over the results of ultrasonic flaw detection. Unfortunately, ultrasonic inspection on land has been abused enough times so that many fabricators and engineers automatically distrust the results.

Like MT, ultrasonic inspection produces no automatic permanent record for later review. Of course there are exceptions to this statement. For instance, ultrasonic machines can be plugged to strip chart recorders, which will record very basic elements of the CRT readout just as photographs or permanent magnetic impressions may be made during magnetic particle inspection tests.

There are some excellent benefits in the use of underwater ultrasonics, however. The seawater environment offers a distinct advantage over ultrasonic testing on land since the surrounding seawater provides an excellent "couplant" to convey the ultra high-frequency sound between the transducer and the material under examination. Thus, one of the major inconsistencies of ultrasonic testing on land, inconsistent or inadequate coupling, is removed in ultrasonic testing underwater. Also, only one side of the material to be examined is required for this test.

The most common method of performing underwater ultrasonic testing utilizes a diver (who often is not a trained ultrasonic technician) to simply move a transducer over the material under examination. The transducer is attached by a long cable to an ultrasonic unit that remains on the surface. A qualified ultrasonic technician viewing the instrument's cathode ray tube (CRT) will verbally "coach" the diver to orient and move the transducer relative to the material under inspection and to any indication areas within the material. This method is rarely satisfactory, however, since it is almost impossible to verbally communicate the fine changes of transducer placement and orientation required to characterize defects.

A newer approach to underwater ultrasonics use is to house a simple ultrasonic thickness-measuring gauge in a watertight housing with a diver-observable direct scale or digital thickness readout. This is satisfactory for uncomplicated thickness measurements where corrosion is known to be general with no pitting. However, a device of this type is ineffective in measuring and characterizing narrow, sharp, or grooved pitting such as will be found in "erosion" corrosion and will give false readings when laminations are present in the material.

It is most important in ultrasonic inspection that the technician performing the test be able to view the cathode ray tube (CRT) of the ultrasonic instrument so that he can move and orient his transducer in such a manner that the sound will be properly introduced into the material as verified by CRT information. It is most important that he observe the CRT information when assessing ultrasonic indications so that he can move his transducer relative to these indications allowing optimization of sound-entry angles, orientation, and measurement of the indications.

Recognizing this, a third approach has been to house a complete ultrasonic testing unit so that the diver may directly observe CRT information while performing the test. This has also proven quite unsuccessful except under the most optimum conditions due to the following problems:

(1) The instrument is usually large and awkward to handle underwater, particularly when a current exists.

(2) The many controls that eminate from the housing may require constant adjustment. The diver is often wearing thick wet suit gloves, which make fine adjustments difficult.

(3) The diver may be faced with trying to hang on to his position underwater, hold on to the instrument, adjust the instrument, and position the transducer – all at the same time.

(4) Poor or "zero" visibility conditions prevent, or at least inhibit, the diver from reading the instrument unless the instrument is pulled directly against his face plate.

(5) The diver may be cold and suffering from lower mental agility due to the influences of depths and his breathing gases. Thus the instrument calibration, adjustment, interpretation of indications, and comparison to specification requirements may be assumed to be detrimentally affected in this environment.

(6) Since the entire inspection is performed solely by the diver, including interpretations of the CRT display, only short lengths of material can be inspected so that the diver can return to the surface to record his inspection results.

Obviously, this approach is far from satisfactory.

The most successful approach to the use of underwater ultrasonics requires that the equipment used:

(1) Is suitable for the underwater environment.

(2) Allows the diver to observe the CRT ultrasonics indications as he performs the test, even in zero visibility conditions.

(3) Is convenient for the diver to operate and will not interfere with his diving requirements and functions.

(4) Will provide a permanent record of the inspection information in a comprehensive method that will allow later analysis under more optimum conditions than those found underwater or on a rocking, noisy boat.

Diver observation of the CRT information in the patented equipment we developed for our own use is accomplished by a television monitor mounted on top of the diver's helmet, arranged so that the video image is beamed by a mirror through a block of plexiglass into the helmet, allowing the diver to view the monitor. The actual ultrasonic instrument is located on the surface with a long cable to the transducer as in previous methods. The CRT information is displayed on the diver's monitor. Thus, the diver, while manipulating the transducer, can observe the resultant CRT indications, allowing him to move and orient the transducer in such a fashion that sound penetration into the material is satisfactory. This allows him to profile, peak, and measure defective areas by reference to transducer movement and the resultant ultrasonic indications.

Also attached to the diver's helmet is a television camera with a wide angle lens oriented so that, water visibility permitting, the transducer location and movement are under observation by the topside console operator. Thus, the topside technician can observe the progress of the inspection including transducer placement and movement. Since the CRT indications and transducer position and movement are observable simultaneously, the technician is able to monitor the quality of the test, the diver's performance, and make intelligent contributions and suggestions to the diver during the course of the inspection.

The topside console displaying all of this information is directly wired to a video tape recorder. The console has the ability to display the CRT indications, the underwater TV camera, and a "split image" of pertinent portions of each image. Thus, the transducer placement and movement can be observed and simultaneously recorded with the CRT ultrasonic indications. Also recorded are the diver and topside technician's comments and observations.

This results in a permanent record of the inspection. The raw data is preserved for later review. Qualified ultrasonic experts other than the original inspectors can review the tapes and confirm, modify, or reject the results and conclusions of the inspection. Since calibration of the equipment is video recorded prior to, during, and after the inspection, the sensitivity, accuracy, and quality of the test are available for consideration during evaluation of the test results. These aspects of a permanent recording make the ultrasonic inspection less dependent upon the ability and judgment of a single individual.

Another method of solving the CRT feedback to the diver for flaw detection has been devised by the U.S. Navy, Panama City, Florida. They have hooked up a signal generator to the CRT screen such that the primary indication upon exceeding a gate level within the instrument will produce an audible sound in the diver's helmet. As the diver manipulates the transducer and peaks up the amplitude of the indication, the sound will correspondingly change. The advantage of this type of system is the use of existing communication facilities to the diver to send the audible CRT information. The drawback is the inability of the diver to perceive the right and left movement of the CRT trace as he moves the transducer. He is also unaware of the other significant CRT indications that may occur on either side of the primary reflector. All of this information is normally of great importance to an ultrasonic technician and aids in the interpretation of the type of defect or reflector he may have. Although the topside technician can view the CRT, the other important input to make a proper evaluation is lacking, specifically the probe orientation to the material under examination.

The Navy is reportedly finding good success in ultrasonic corrosion characterization with a computer-based system. The diver has no feedback and simply "massages" the material with the transducer. The computer samples the readings observed, perhaps hundreds of times per second, makes a statistical analysis of the information, and provides a numerical description of the material condition.

Remote Controlled Underwater Vehicles (RCV)

The use of remote controlled underwater vehicles to perform underwater nondestructive tests is very attractive. Several major advantages may be enjoyed by use of this piece of equipment, as follows:

(1) The financial savings in using this kind of system as opposed to a full-blown diving operation. This is particularly significant in depths of 300 feet and greater where the cost of the diving operations is astronomical.

(2) The inherent safety advantage of having a machine under the water rather than a man.

(3) The ability to operate in more difficult weather conditions than would normally be allowed for a diving operation.

To date, RCV vehicles have been used for underwater visual inspection and to some extent for underwater potential measurements. Underwater ultrasonic inspection by RCV's has not yet been satisfactorily developed. Although the problem sounds simple, the execution is very difficult. Securing these vehicles to a platform or pipeline, cleaning the area to be examined of marine growth, and positioning the ultrasonic transducer satisfactorily are the main areas of difficulty. Nevertheless, it is the author's opinion that within five years underwater ultrasonics via RCV's should be a standard method.

The disadvantages of RCV's are the difficulty in maneuvering in turbulent water conditions and strong currents, and their becoming entangled or "lost" in complicated underwater structures. Work is being performed in all of these areas, however, and these problems should one day be solved. It is the author's opinion that RCV's will one day play a major role in many underwater nondestructive inspections.

Conclusion

It is clear that the foregoing methods of nondestructive testing face many problems and difficulties when used underwater. A good understanding of the strengths and weaknesses of each type of test, when used underwater, is fundamental to obtain the information needed to make reasonable engineering and maintenance decisions for underwater structures.

Bibliography

R. Frank Busby Associates. 1978. *Underwater inspection/testing/monitoring of offshore structures.*

Krolicki, R.P. 1977. *Internal corrosion examination and wall thickness measurement of pipe by radiographic method.*

Cathodic protection fights corrosion. *Offshore Magazine* July 1979.

Underwater nondestructive testing equipment and techniques. Barrett, Fred, and Mittleman, John. Presented at the International Diving Symposium, New Orleans, Louisiana, February 1979.

AWS D3b Recommended Practices

E.A. Silva
Office of Naval Research

Historical Overview

The AWS D3b Subcommittee on Underwater Welding and Cutting had its first meeting on 7 April 1975. The group was formed in response to perceptions that underwater welding had matured to a state where welder/divers could produce welds of sufficient quality to consistently meet the requirements of a wide variety of offshore users; that there was sufficient variability in the quality of underwater welds to require a means for ensuring that proper joining was done; and that the existence of a formal code or specification would make potential users more likely to select underwater welding as a routine joining technique.

The D3b Subcommittee was formed with roughly equal representation from three major groups: the "Doers," those who make underwater welds; the "Users," those who would be customers for this service; and the "Regulators," those who could be or were involved in the subject through Federal, classification society, or academic auspices. International interest in the activities of the Subcommittee was apparent early, and participating and advisory members from the United Kingdom and the Federal Republic of Germany joined the group. The Chairman also represented D3b at various International Institute of Welding gatherings.

As work progressed, it became clear that underwater cutting offered less opportunity for specification development than welding. Therefore, the D3b Subcommittee devoted the majority of their efforts to underwater welding and elected to defer work on underwater cutting.

The Subcommittee met for five years. During this period, membership changed to reflect new interest in the field, individual points of view shifted, and, most interestingly, underwater welding technology improved and was used more frequently as the group went about its business. The product of these efforts was a final draft of the AWS Specification on Underwater Welding. It was presented to the AWS D3 Committee on Welding in Marine Construction on 16 April 1980.

Initial Concepts

There are five general types of underwater welding being used commercially:

(1) Welding in a pressure vessel where the pressure is kept at approximately one atmosphere: "one atmosphere welding"

(2) Welding in a chamber from which the water has been displaced at ambient pressure in an atmosphere that does not require diving equipment: "habitat welding"

(3) Welding in a chamber that accommodates the head and shoulders of the welder/diver who is in full diving dress: "dry chamber welding"

(4) Welding at ambient pressure in a small enclosure with the welder/diver in the water: "dry spot welding"

(5) Welding at ambient pressure with the welder/diver in the water and without a physical barrier between the water and the welding arc: "wet welding"

The Subcommittee recognized that there are other procedures that cross these definitions, and no attempt was made to exclude other processes. Consideration of these general types allowed the Subcommittee to focus on the unique effects of the underwater environment on welding and to develop a series of special quality requirements and inspection procedures to ensure proper underwater welds.

The goal of the resulting Specification is to define the important variables and describe welding and inspection procedures so that bonds of known quality can be specified for the carbon and low alloy steels commonly used. Gas metal arc, gas tungsten arc, flux cored arc, plasma arc, and shielded metal arc welding are considered the primary processes of interest. However, other processes can be used if they can meet the qualification requirements of the Specification. They must also be acceptable to the authorized representative of the owner or engineering agency in charge of construction (referred to hereafter as the "customer").

Weld Types

A series of four weld types have been defined to establish a level of serviceability and a set of required properties. The customer specifies the weld type and can specify additional requirements to meet an individual need. Type A welds are intended for structural applications and are suitable for use with the same design stresses as their above-water counterparts. Type B welds are intended for limited structural applications, and Type C bonds are available for situations where load bearing is not a primary function.

The final type of weld is called Type O. It is the underwater equivalent of welds made at the surface in compliance with a code or standard developed for the particular construction involved. In this case, the provisions of the in-air specifications are met, plus those required to ensure acceptable underwater service.

Workmanship

As would be expected, the appearance and size of underwater welds should be what was intended by the joint designer. Some special concerns related to underwater welding includes limiting the coatings (including waterproofing) on consumables to those used for qualification, and the specification of methods for the transport and storage of filler metals in the welding procedure.

Base metal preparation underwater includes the removal of all paint, marine growth, and other foreign matter. Fit-up requirements for dry welding are slightly different in minimums than those for in-air procedures. They reflect a need for ready access. For cases involving other than dry welding, the tolerances developed and approved during procedure qualification are used. An important feature is the requirement to monitor current, voltage, shielding gas composition, and flow rate at the underwater welding worksite.

Tack welds are subject to the same procedure and quality requirements as the final welds, and temporary welds must conform to the requirements of the specified weld type.

A unique requirement for underwater welding is that of a "confirmation weld." This is a test weld made at the underwater worksite on a material of similar carbon equivalent and thickness as the production material (but need not be thicker than 12 mm [1/2 in.]). The purpose of this weld is to confirm that the welding system is operating correctly. Only one confirmation weld 150 mm (6 in.) long is required in the most difficult position. The weld examination requirements should be the same as for the production weld and are never less than visual examination.

The visual examination requirements for Type A welds are comparable with those given in AWS D1.1, Structural Welding Code – Steel. No cracks, surface porosity, or entrapped slag are allowed. Undercut is limited to 0.8 mm (1/32 in.) in depth and cannot exceed 50 mm (2 in.) in a length of 0.4 mm (1/64 in.) undercut in any 30 cm (12 in.) of continuous weld. Type B welds have profile requirements less stringent than those for Type A welds and can have no cracks. Visible piping porosity is allowed as long as it does not exceed 1.6 mm (1/16 in.) in diameter and does not exceed 9.5 mm (3/8 in.) in any linear 25 mm (1 in.) of weld.

Type C weld visual examination requirements are similar to Type B, except that piping porosity is not restricted and reinforcement of groove welds is limited to the greater of 20 percent of the base metal thickness or 5 mm (3/16 in.). Type O welds must meet the requirements of the controlling terrestrial code.

Type A welds are to show no cracks under radiographic examination while Type B welds are allowed crater cracks with their greatest dimension less than 4 mm (5/32 in.). Pores smaller than 1.6 mm (1/16 in.) in diameter are not restricted in number, and isolated slag inclusions up to 3.2 mm (1/8 in.) in width are allowed for the Type B case. Type C welds are not restricted in soundness when examined by radiography except that weldments can contain no cracks other than crater cracks up to 4 mm (5/32 in.) long. As would be expected, Type O welds must meet the radiographic examination requirements of the specified code.

Procedure Qualification

Both the procedure and the welder must qualify to meet the proposed specification. A procedure can be qualified by testing weldments made under actual or simulated site conditions. As is typical, essential and nonessential variables are listed, and a change in an essential variable beyond the conditions specified requires development of a separate procedure.

Some examples of essential variables that reflect underwater welding requirements are changing the procedure for filler metal underwater transport and storage and a decrease in accessability or visibility from a change in the dimensions of a shroud or component. Similarly, change in weld back-side from dry to wet or vice versa is an essential variable for some processes, as is an increase in displacement gas oxygen partial pressure or a change in displacement gas composition beyond that associated with a specified depth change allowance. Introduction of a new gas into a previously qualified composition, addition or deletion of supplementary coatings or waterproofing, changes in any type of barrier to restrict water access during welding or cooling are all essential underwater welding variables.

Nonessential variables related to the wet environment include a change in diving mode, the presence or absence of water currents, and a change from fresh to salt water, or vice versa.

Procedure qualification begins with knowing the chemical composition and carbon equivalent of the test plate. This is obtained by chemical analysis. Composition of the workpieces can be determined by using the maximum values from the certified range on mill certificates; maximum values from specifications, if all the production material can be shown to be produced in accordance with these specifications; maximum values obtained during production of the material; and from the highest values in historical data, if it can be shown that the data includes all the materials to be welded in production.

Carbon equivalent has been found to be a good means for predicting underwater weldability of steels. The formula selected for inclusion in the specification was, in weight percent:

$$CE = C + \frac{Mn}{6} + \frac{Cr + Mo + V}{5} + \frac{Ni + Cu}{15}$$

or:

$$CE = C + \frac{Mn}{6} + 0.05$$

when only carbon and manganese are known.

The carbon content of the test pieces must be within 0.10 percent of the production material and substantial amounts of eleven other alloy materials, such as copper, nickel, cerium, sulfur, or titanium, require a similar composition in the procedure test plate. This, of course, affects the range of qualification obtained per test.

Depth considerations are a unique concern of the underwater welder, and the maximum depth permitted for production welding is limited to the procedure qualification depth plus 10 m (33 ft) or 20 percent of the procedure qualification depth, whichever is greater. In order to avoid unnecessary procedure qualification at a great number of steps, it was agreed that qualification at two depths qualifies for all depths in between.

The minimum depth permitted for production welding is equal to the procedure qualification depth less 10 m (33 ft) or 20 percent of the procedure qualification depth, whichever is shallower. An exception to this is that if welds are to be made from the surface to 3 m (0 to 10 ft), they must be qualified at the depth of the production welds or shallower. This is because shallow underwater welding has been found to differ substantially from deeper welding performance. Special notice to beware of wave effects at shallow depths is also given.

Position is an essential variable for underwater welding. The positions to be welded in production determine the positions required for groove welding Type A test plates (Figs. 1, 2, and 3). Fewer tests are required for Type B procedure qualification, and Fig. 4 and 5 illustrate the appropriate test pieces. Type C groove weld specimens are limited to the four macroetch specimens shown in Fig. 6.

Type O groove weld test assemblies must meet the dimensional requirements of the customer. If the base metal on both sides of the weld required impact testing, the weld metal and heat-affected zone in the procedure qualification specimen must be impact tested. There is also a requirement for an all-weld-metal tensile test.

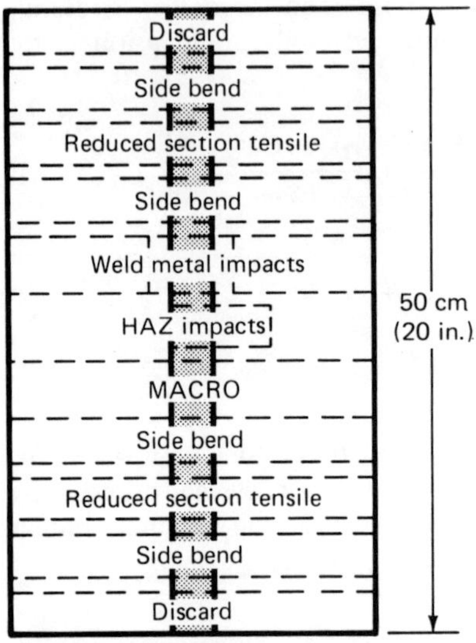

Location of test specimens on welded test plate over 18 mm (3/4 in.) thick; first position

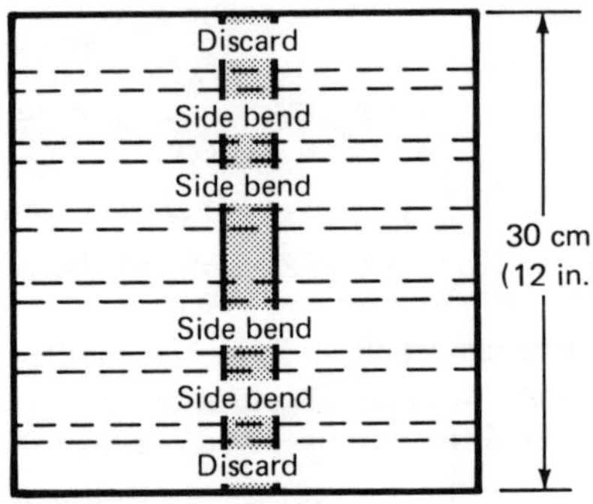

Location of test specimens on welded test plate over 18 mm thick; additional positions

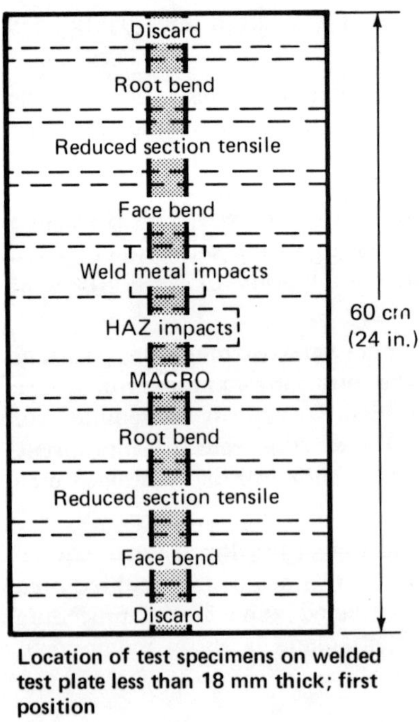

Location of test specimens on welded test plate less than 18 mm thick; first position

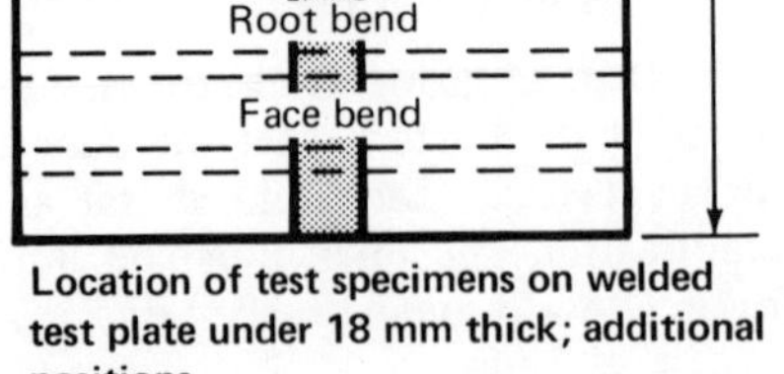

Location of test specimens on welded test plate under 18 mm thick; additional positions

Fig. 1 – Type A test plates

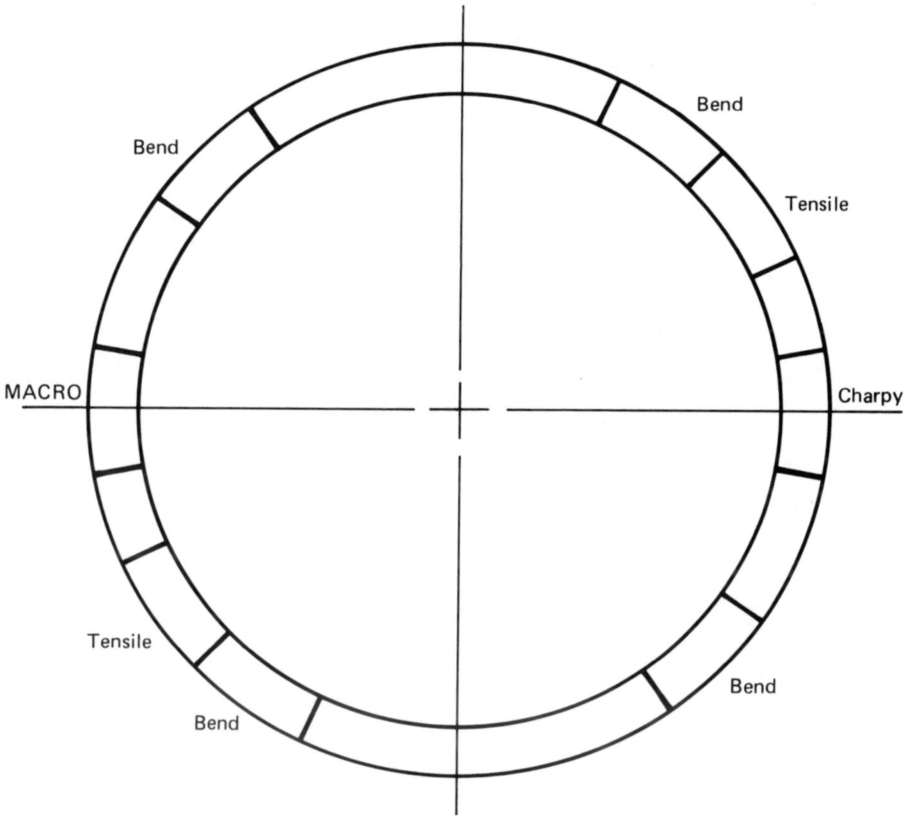

Location of test specimens for welded test pipe

t < 18 mm (3/4 in.) two root bends and two face bends

t ⩾ 18 mm four side bends

If the pipe diameter is too small to permit all specimens to be machined from a single weld, additional welds shall be made to provide sufficient material for each test

Fig. 2 – Type A pipe test specimen

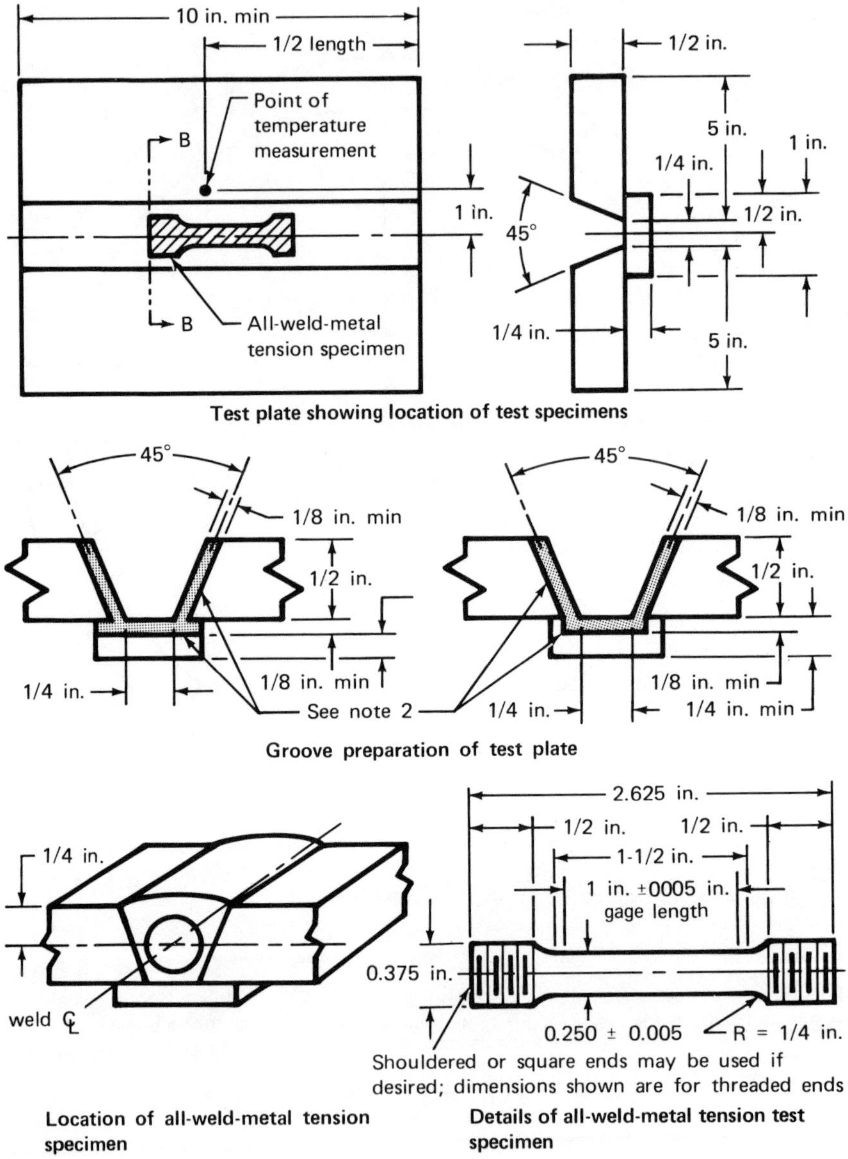

Notes:

1. Either one of the backings shown in (b) below may be used to test any diameter electrode.
2. Edges of the grooves and the contacting face of the backing shall be surfaced as shown by any size of the electrode being tested before welding.
3. Each layer shall be approximately 1/8 in. thick.

Fig. 3 – All-weld-metal tensile and Charpy specimen

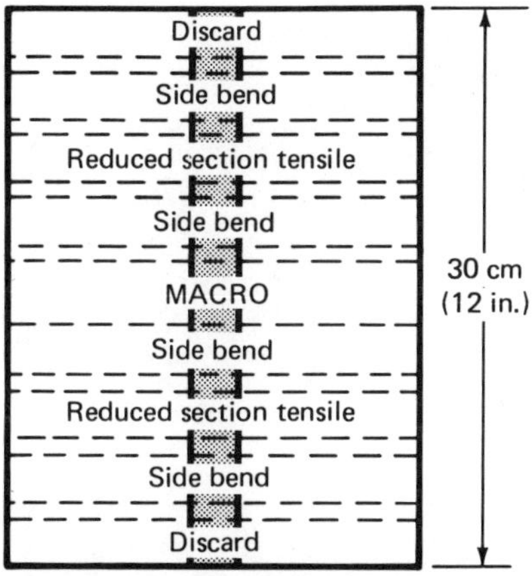

Location of test specimens on welded test plate over 18 mm thick; first position

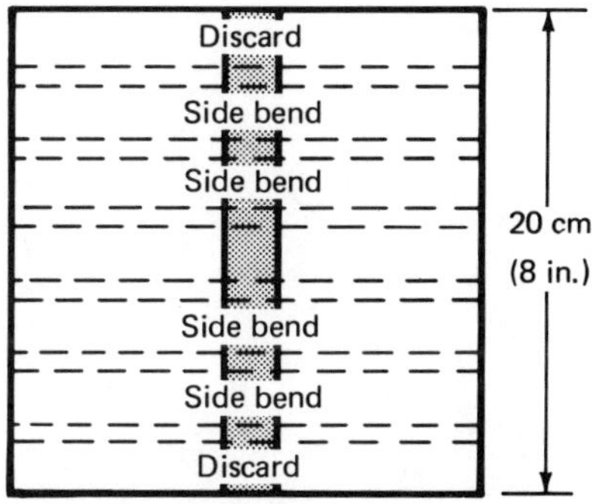

Location of test specimens on welded test plate over 18 mm thick; additional positions

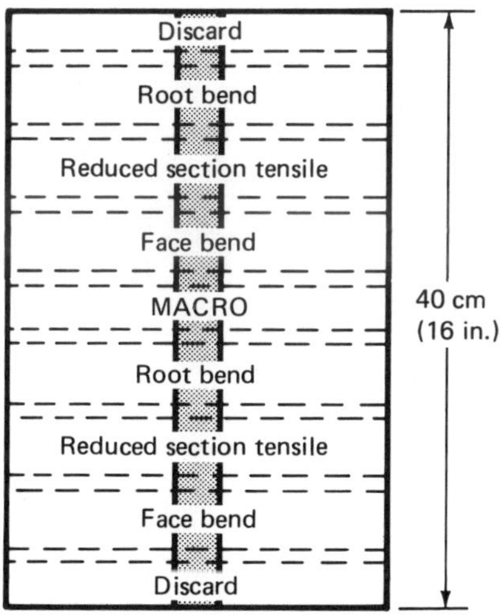

Location of test specimens on welded test plate less than 18 mm thick; first position

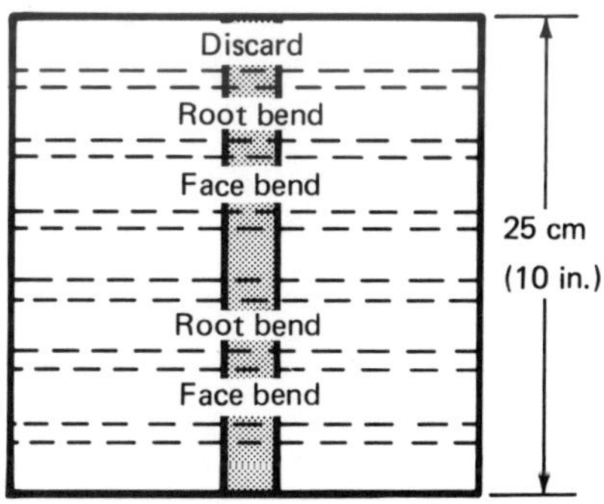

Location of test specimens on welded test plate under 18 mm thick; additional positions

Fig. 4 – Type B test plates

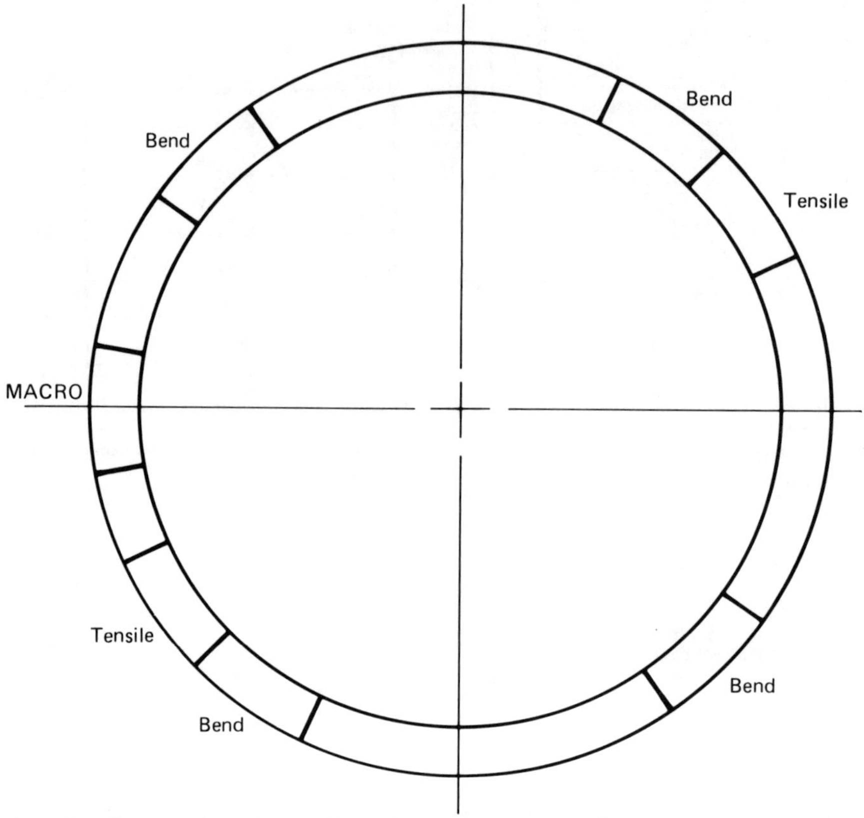

Location of test specimens on welded test plate under 18 mm (3/4 in.) thick; first position

Fig. 5 – Type B pipe test specimen

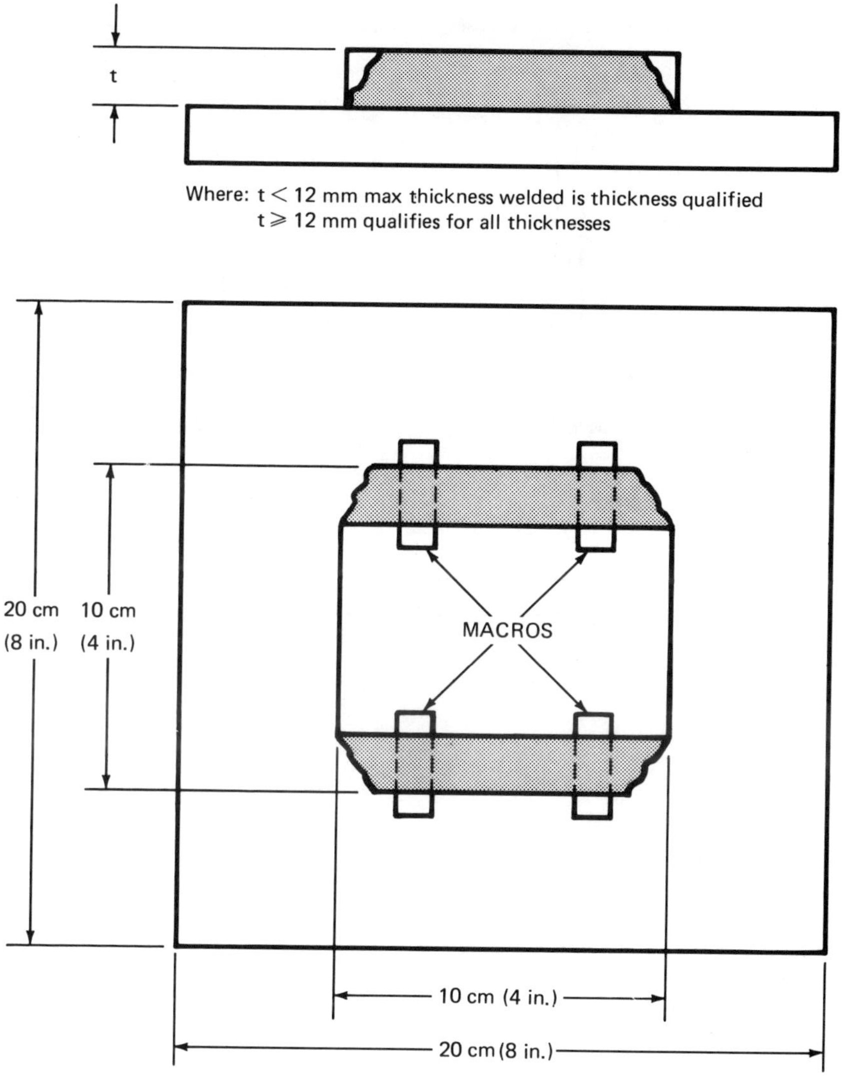

Fig. 6 – Type C weld specimen

Type A fillet weld procedures are qualified by producing an all-weld-metal tensile specimen (Fig. 3) with sufficient length for Charpy impact testing of the weld metal and two fillet weld tensile specimens, as illustrated in Fig. 7, plus a fillet weld break test specimen described in Fig. 8.

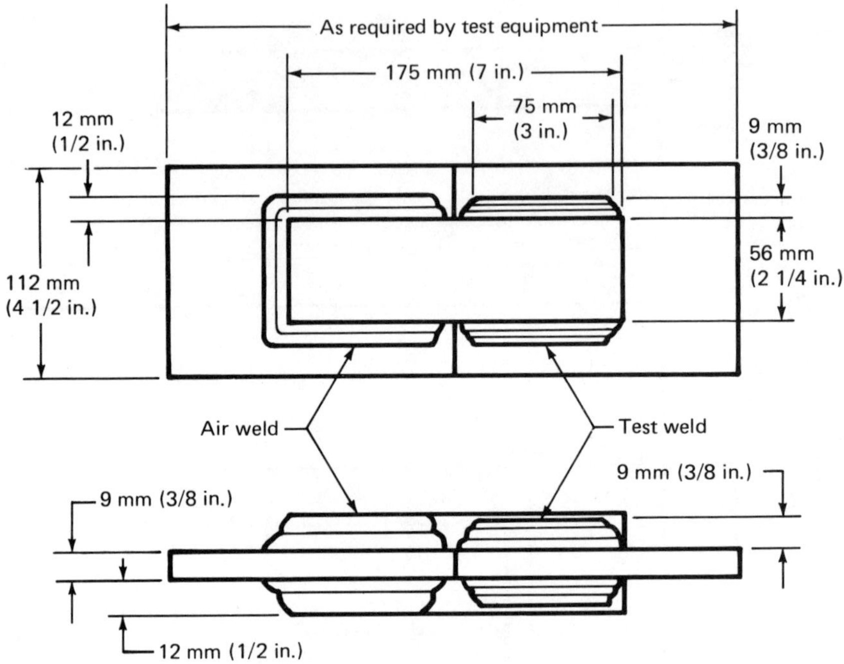

Fig. 7 – Fillet weld tensile specimen

Type B fillet welds require the two fillet weld tensile specimens and a fillet-weld-break specimen. Both the Type A and B fillets can be addressed by welding groove weld assemblies; plus a fillet weld fracture specimen for the Type B case; plus an all-weld-metal tensile specimen for the Type A case.

Type C fillet weld procedures are qualified with the same four macroetch specimens used for the groove case. These requirements are presented in greater detail in Table 1.

Testing of the specimens is routine with the exception that Vickers hardness testing is required on the macroetch specimens for Type A welds. The microhardness measurements cannot exceed 325 VHN with a 10 kg load. This addresses the quenching effects associated with underwater welding.

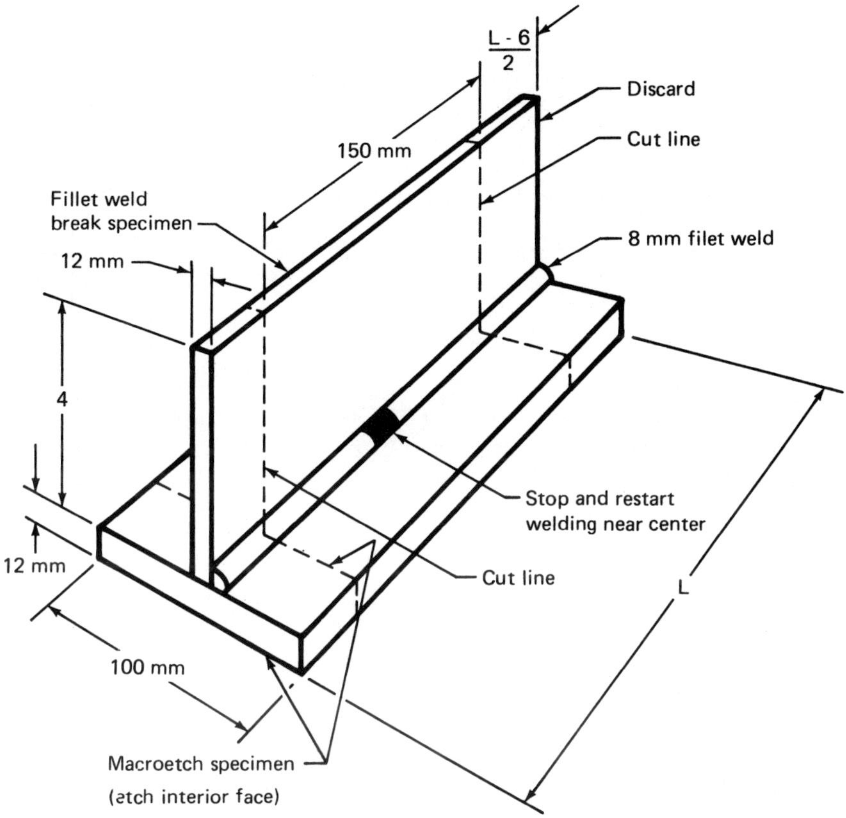

Note: Plate thickness and dimensions are minimum.

Fig. 8 – Fillet-weld-break and macroetch test plate

Table 1
Number and type of test specimens – procedure qualification

Coupon	Weld	Thickness	Visual	Radio-graph	Red. sec. tensile	Fillet weld shear	All-weld-metal tensile	Bends, roots, face	Slide	Macro	Charpy	Fillet-weld-break
Plate	Groove A	< 12 mm (1/2 inch)	X	X	2	–	1	4	–	1	WM + 11AZ	–
		⩾ 12 mm[a]	X	X	2	–	1	–	4	1	WM + 11AZ	–
Pipe	Groove A	< 12 mm	X	X	2	–	1	4	–	1	WM + 11AZ	–
		⩾ 12 mm[a]	X	X	2	–	1	–	4	1	WM + 11AZ	–
Plate	Groove B	< 12 mm	X	X	2	–	–	4	–	1	WM + 11AZ	–
		⩾ 12 mm[a]	X	X	2	–	–	–	4	1	–	–
Pipe	Groove B	< 12 mm	X	X	2	–	–	4	–	1	–	–
		⩾ 12 mm[a]	X	X	2	–	–	–	4	1	–	–
Plate	Fillet A	< 12 mm[d]	X	–	–	2	1	–	–	2	WM	1
	or	< 12 mm[e]	X	X	2	–	1	4	–	3	WM + 11AZ	1
Plate	Fillet B	< 12 mm	X	–	–	2	–	–	–	2	–	1
	or	< 12 mm	X	X	2	–	–	4	–	3	–	1
Plate	Fillet C	18 mm (3/4 inch)	X	–	–	–	–	–	–	4	–	–
Plate	Fillet O	18 mm	X	b	c	–	1	c	c	1	c	c

a. Qualifies fillet welds where throat exceeds 12 mm.
b. Groove welds are radiographed.
c. Mechanical test requirements shall satisfy the service specific code referenced by the customer. Weld metal and heat-affected zone Charpy toughness testing is required if the base metal on both sides of the weld must meet an impact requirement.
d. Fillet weld tensile option.
e. Groove weld tensile option.

Elongation criteria range from at least 12 percent for 620 MPa (90 ksi) steel to 19 percent for 345 MPa (50 ksi) material for the A, B, and O types. After discarding the highest and lowest values, the Charpy impact results must be in accordance with the following limits:

Base metal specified minimum tensile strength	Ave/Min Charpy-V
to 485 MPa (to 70 ksi)	20/14J (15/10 ft · lb)
486 to 550 MPa (71 to 80 ksi)	27/19J (20/14 ft · lb)
551 to 690 MPa (81 to 100 ksi)	34/24J (25/17 ft · lb)

Type C welds need only meet the visual quality requirements for Type B welds and not show cracks under 5X magnification on the macro-etched specimens.

Type O welds must meet the requirements of the code selected by the customer plus an all-weld-metal tensile test that meets the following requirements: to 345 MPa (50 ksi) yield, 19 percent elongation; to 620 MPa (90 ksi) yield, 14 percent elongation; 620 MPa yield or higher, 12 percent elongation.

The proposed specification allows the customer to specify supplemental requirements for welds of any type. The goal here was to provide a minimum – not to restrict the test types or methods.

Welder Qualification

Welder qualification tests are intended to establish a welder's ability to make sound welds using a qualified underwater welding process. They can be done under simulated or actual production conditions provided that the essential variables for the qualification are satisfied. The need for records of qualification is stressed in the Specification.

If the candidate fails to qualify in a given position, he or she may be retested in that position by making two test welds. Both welds must be acceptable, or evidence of training is required prior to another retest.

Qualifications hold for a year from the last qualification or acceptable production weld. However, a 200 mm (8 in.) weld in 12 mm (1/2 in.) thick material that conforms to visual and radiographic weld requirements must be made in the most difficult position to be welded in production if a welder has not worked with the procedure during the previous ninety days.

Diving considerations are covered by constraints for visibility and support similar to those provided during qualification. Also the thermal environment and length of exposure cannot be substantially more severe than that experienced during qualification.

Essential variables that relate specifically to underwater welding include exposure to water, waterproofing of the electrodes, shielding gas, removal of backing material, a breathing gas change from mixed gas to air, depth effects, a change from sea water to fresh water for wet welding, and a change from hot rolled or normalized steels to quenched and tempered material.

Type and position limitations are similar for the underwater cases to those for terrestrial welding with the addition of specimens for pipe fillet welding. A weld on a 6GR test assembly is required to qualify a welder for complete joint penetration groove welds in tubular T-, Y-, or K-connections. Testing on a 12 mm (1/2 in.) or greater wall thickness qualifies for all wall thicknesses and diameters.

A fillet weld break and macroetch test plate are required for Type A and B welds. Type C qualifications are done with the procedure qualification coupon. Groove T-, K-, and Y-welds are required to meet the visual and radiographic quality standards for a given weld type. Ultrasonic inspection and macroetched specimens can be used in place of radiographic inspection. Fillet welds must meet the procedure qualification standards for a given weld type.

Inspection of Underwater Welds

Type A welds must be visually examined and full penetration welds must be either radiographed or ultrasonically inspected. Partial penetration and fillet Type A welds must be examined with magnetic particle techniques. Because the Inspector may not have access to the underwater worksite, special consideration for the collection and recording of evidence is generally necessary. This can take a wide variety of forms including photography, television tapes, and extensive documentation.

The most common underwater examination techniques are visual, radiography, ultrasonic, and magnetic particle techniques. Dye penetrant examination procedures suitable for hyperbaric use may also be used in a dry environment. However, this method has not been found suitable for examination of welds in most wet situations.

Visual inspection underwater is similar to that in air. However, the effects of artificial lighting and the water require that more attention be given to the use of gauges and aids to demonstrate compliance. Many other procedures for examination are quite similar to terrestrial methods. For example, in the dry hyperbaric cases, ASTM Specification E 109 (Dry Powder) is used for magnetic particle inspection. For the wet case, the wet powder specification, E 138, Standard Method for Wet Magnetic Particle Inspection, is prescribed.

Radiographic examination is a common underwater approach. The Specification allows the customer to consider exceptions to normal practice, such as radiographic examination of fillet, T-, or corner welds or changes in radiographic procedures, as in AWS D1.1, Structural Welding Code – Steel. A written procedure specifying the equipment and operations required must be approved by the customer's inspector prior to production recording. Supporting data to show the resolution expected must be attached to the procedure. Underwater radiographic procedure variables are the same as those in air.

Ultrasonic examination techniques are readily applied underwater, and the Specification contains a detailed section addressing this important technique. Ultrasonic procedures must be prepared or approved by an individual qualified as an ASNT-TC-1A Level III who is experienced in the ultrasonic examination of underwater welded structures. This is in keeping the specification guidance that nondestructive examination personnel should be ASNT Recommended Practice Number SNT-TC-1A or equivalent qualified. They should also have evidence of training and experience in the operation of equipment and performing appropriate evaluations at marine job sites.

Conclusion

This completes the overview of the proposed AWS Specification on Underwater Welding. A few remarks on some of the philosophical issues it raises deserve mention. What we have here is the product of a labor that attempted to bring together representatives of those involved with all aspects of underwater welding, from users to researchers, to document their best insights. It was recognized that producing an underwater welding code would result in increased costs for underwater welding firms because of items such as qualification dives and record keeping. Also, it might provide avenues to lessen the competitive edge gained by a few of the firms that have a lead in this field. However, there was also a need to illustrate that underwater welding was no longer a technique to be used for temporary or odd jobs. It can be done well, and entirely adequate bonds can be made in the marine environment for both permanent repair and construction. The existence of a specification that will allow designers to call out underwater welds will help remove them from the unusual to a routine welding alternative. This will bc good for our industry and provide the design community with greater flexibility in their work.

It is possible that Class O welds will not be required in the future. AWS underwater welds, as defined in future editions of this Specification, will be recognized as meeting the needs of the offshore community and regulators. This will come with time and good experience with the Specification.

It should be restated that no attempt was made to limit welding processes and approaches. New processes came and went during the writing of the document and surely more will follow. By working with weld quality, rather than specific techniques, the opportunity is open for any method to be applied. The issue is not how underwater welds are made, but whether they can be made to the standard of the Type for which they hope to qualify.

Underwater welding is growing into a routine construction practice. The AWS Specification on Underwater Welding is an attempt to provide guidance to protect the welding and construction industries while this growth occurs. It is a major step toward underwater welding respectability.

Problems Encountered in Underwater Welding: Facilities Available for the Study of Certain Parameters in a Simulation Chamber

J-P. Gaudin
Institut de Soudure
and
B.G. Sudreau
TOTAL-Compagnie Francaise des Petroles

Introduction

Welding safely under hyperbaric conditions has been the objective of TOTAL - Compagnie Francaise des Petroles for the past 10 years. For the construction of the FRIGG pipeline in the North Sea, as early as the engineering stage, TOTAL-CFP asked the Institut de Soudure to investigate the problems raised by underwater welding, particularly under hyperbaric conditions, in order to develop the specifications and procedures to be used for the jointing of subsea pipelines.

In hyperbaric welding the action of pressure, which is a significant thermodynamic parameter, requires that certain concepts commonly accepted at atmospheric pressure be reconsidered and taken into account when developing welding specifications with a view to the qualification of a welding procedure.

But welding safely cannot be achieved without previously investigating a certain number of factors inherent to welding. In view of the costs of tests under hyperbaric conditions involving the presence of man, the use of sufficiently sophisticated simulation equipment enables the investigation of problems raised by the effect of pressure up to the stage of prequalification of a welding procedure. The thorough knowledge of the techniques to be used will make them operational with a maximum guarantee.

What is Underwater Welding?

This question may seem useless for the specialist, but it deserves being discussed whenever welding is examined in terms of safely producing a joint. Underwater welding can be defined either in terms of the welder's situation during welding, or (Fig. 1) by considering the process itself, regardless of the welder. In this case, the definition is easy since only one possibility remains: hyperbaric welding. The processes differ only by the size of the chamber where welding takes place, water thus being only a variable among many others that may possibly affect the evaluation of the considered process.

This is the only definition of hyperbaric welding that will be considered in this paper. Nevertheless, since we are dealing with deep sea welding, hyperbaric welding will mean dry welding in a chamber at the hydrostatic pressure of the particular depth.

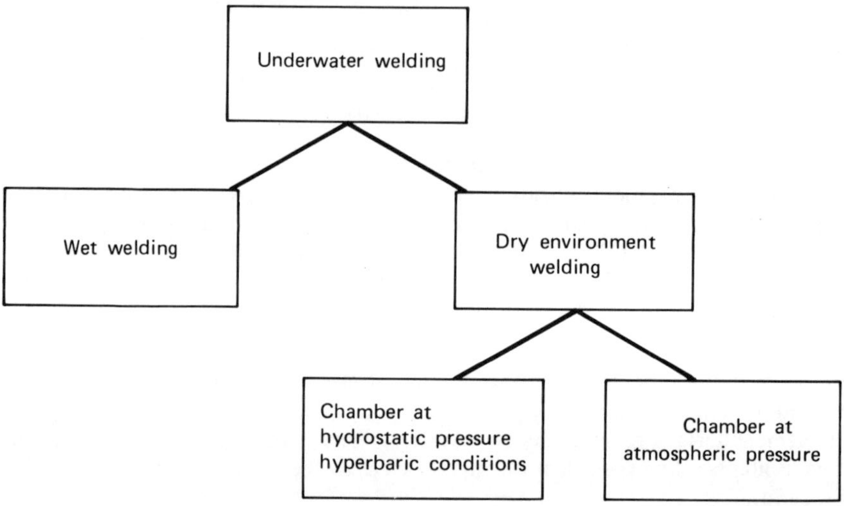

Fig. 1 – Different possibilities of underwater welding in terms of welder's situation

Factors Affecting Hyperbaric Welding and Their Effect on the Execution of Welds

In hyperbaric welding, as in welding at atmospheric pressure, influencing factors are physical, chemical, and metallurgical; they are affected only by pressure and by the nature of the environment. Therefore, we will try to determine their influence both on the welding operation and on the quality of the joints thus produced.

Health and Safety. For the welder, health and safety depend on the environment and on the selected diving technique.

Although this paper does not aim to deal with welder physiology, this point cannot be neglected because of its significant effects on the choice of the welding techniques.

In wet welding, safety consists mainly of protecting the welder against stray currents that are liable to make his work dangerous.

In hyperbaric welding the welder works in a gas, thus making electrical hazards secondary; the physiological aspect may therefore prevail; hence, two possibilities may be considered.

The welder is in the same gas as the area around the weld, without a special breathing protection. Although this technique has not yet been used in spite of the successful results of the first tests, we cannot start this section without mentioning WELDAP welding conducted on the seabed at atmospheric pressure.

Things are, however, totally different when it comes to welding in a hyperbaric chamber. Let us recall that the pressure inside the chamber is equal to the hydrostatic pressure of the particular depth. The use of air is limited by the partial pressure of oxygen, and it is often admitted that a depth of 70 meters is the limit beyond which there is a risk of oxygen enrichment together with an alteration of the lungs and the nervous system. On the other hand, pressure intensifies the noxious effects of carbon monoxide, carbon dioxide, and ozone on the body. Not all gases, if breathed, are harmless to man. Moreover, let us recall that certain so-called "neutral" gases such as argon cannot be considered as such under hyperbaric conditions. If we add the harmful effects of the inhalation of mineral dusts, it becomes easy to understand why gases or foreign elements may be introduced into the welder's environment only if TLV's are taken into account (Fig. 2).

Different analyses of the environment made during welding have shown the influence of pressure on the pollution of the chamber by the elements that are most likely to be encountered in welding, i.e., carbon monoxide, carbon dioxide, particles of metal oxides in suspension, etc. (Figs. 3 and 4).

Elements	Concentration
CO_2	5,000 ppm
CO	50 ppm
NO_2	5 ppm
O_3	0.1 ppm
Iron (oxide fumes)	5 mg/m^3
Mg (oxide fumes)	10 mg/m^3
Fumes	10 mg/m^3

Fig. 2 – TLV for several gases and fumes at atmospheric pressure

Besides this increase in pollution of the welder's environment caused by the confinement of the atmosphere he breathes, it must be added that the thermal conductivity of helium, which is commonly used as an environment in hyperbaric chambers, is 7 times as high as that of air, thus likely to affect the thermal equilibrium of the human body.

If the welder is to work in optimal conditions, all these considerations must be taken into account, as are the welding variables.

The welder works in a gas differing from the one he breathes. Regardless of the thermal conductivity of the gas and of the thermal equilibrium of the body, the pollution problems in the chamber are not as crucial as in the above case. Since the welder cannot breathe this gas, he must wear a breathing mask, which protects him against intoxication. Here the problems concern the breathing equipment, which must be perfectly leaktight and not liable to be damaged during welding.

Operating Variables and Metallurgical Factors. To simplify the discussion of the different physical, chemical, and metallurgical factors affecting welding and to take them better into account when examining their influence on the qualification of welders and procedures and on the properties of the joints thus produced, these factors were voluntarily divided into two groups:

(1) Operating variables governing the welding conditions

(2) Metallurgical factors, sometimes resulting from the operating variables and determining the quality of the welded joints

Environment	Chamber		Arc	Concentration					
	Type	Volume	time	O_3 ppm	NO_2 ppm	CO ppm	CO_2 %	HF ppm	Fumes mg/m^3
Helium	Simulation chamber	70 L	107 s	0	0	500	< 0.01	0	34.5
	Actual manned chamber	20 m^3	2 x 4 h			470			32.5
Helium + O_2 (400 mb)	Simulation chamber	70 L	105 s	Traces	0	210	< 0.02	0	58.7
	Actual manned chamber	20 m^3	2 x 4 h			201			55.3
Helium + H_2O (80% hr)	Simulation chamber	70 L	108 s		0	500	0.02	0	53.2
	Actual manned chamber	20 m^3	2 x 4 h			470			49.6
Helium + O_2 (400 mb) + H_2O (80% hr)	Simulation chamber	70 L	53 s	0	0	200	0.01	0	17.7
	Actual manned chamber	20 m^3	2 x 4 h			380			33.7
T L V				0.1	5	50	0.5		10

Fig. 3 – Concentration of gaseous contaminants in shielded metal arc welding (SMAW)

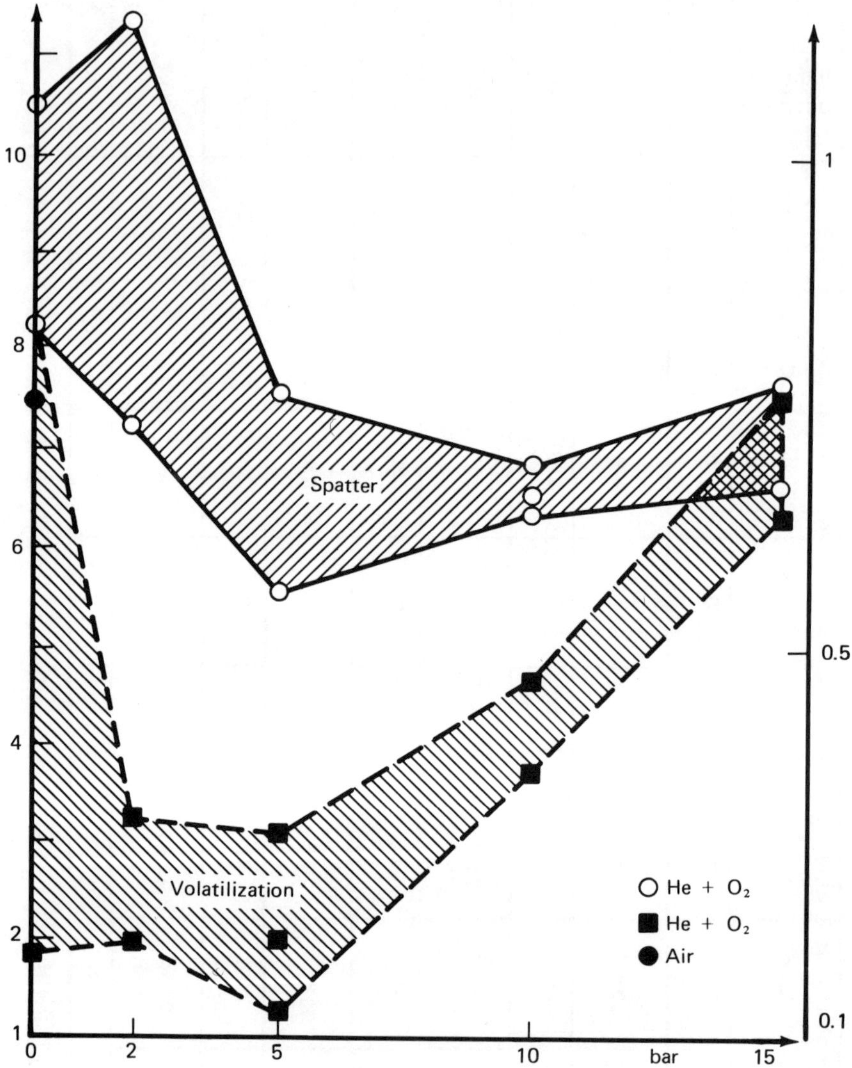

Fig. 4 — Losses by spatter or volatilization (per 100 g melted electrode)

Operating variables. It is practically impossible to list all factors affecting welding because they have numerous origins. This is even more difficult to do for manual welding where man plays an important part hard to characterize in terms of measurable variables that may subsequently be controlled. Therefore, on the basis of the processes that are commonly used in hyperbaric welding, we will try to show some aspects of the influence of pressure on these operating variables.

Shielded metal arc welding. Pressure modifies the characteristics of the electric arc, and it is now well known that the electrical variables of welding are thus affected by a modification of the physical properties of the plasma. It is therefore interesting to examine the effects this has on the geometry of the weld bead thus made. This study was conducted according to certain dimensional criteria (Fig. 5) on beads made in the flat position with E7018 electrodes. It shows that the geometry of the bead is very much altered by pressure, in particular in the range corresponding to a depth of 50 m (5 bars).

In vertical-up welding, test results show that the electrode diameter as well as the type of covering are important factors that may increase the previously noted changes in bead geometry.

Thus, contrary to what happens at atmospheric pressure, for a same class of electrodes, e.g., E7018, the composition of the covering becomes an essential variable that may condition the usability of the electrode much more than at atmospheric pressure.

On the other hand, it appears that raising the arc temperature increases the fluidity of the molten pool and alters surface tensions, thus making electrode diameters above 3.15 mm (1/8 in.) practically useless. Selecting the proper electrode diameters is a determinant in avoiding defects such as porosity and slag inclusions.

Without listing all characteristics of a welding power source, let us briefly recall what these properties should be in relation to operating conditions in manual welding. A power source should ensure a satisfactory arc initiation, ensure reignition of the arc after each short circuit, ensure a good regulation of arc length, and have minimum power variation.

On the basis of the qualification of a welding power source, it is interesting to determine the effect of pressure on metal transfer in the arc and to try to examine the dynamic behavior of the different power sources.

This was done by testing three of the most common types of commercially available power sources:

(1) "Drooping" power source of former design

(2) Transistorized "vertical" power source

(3) Chopper type power source

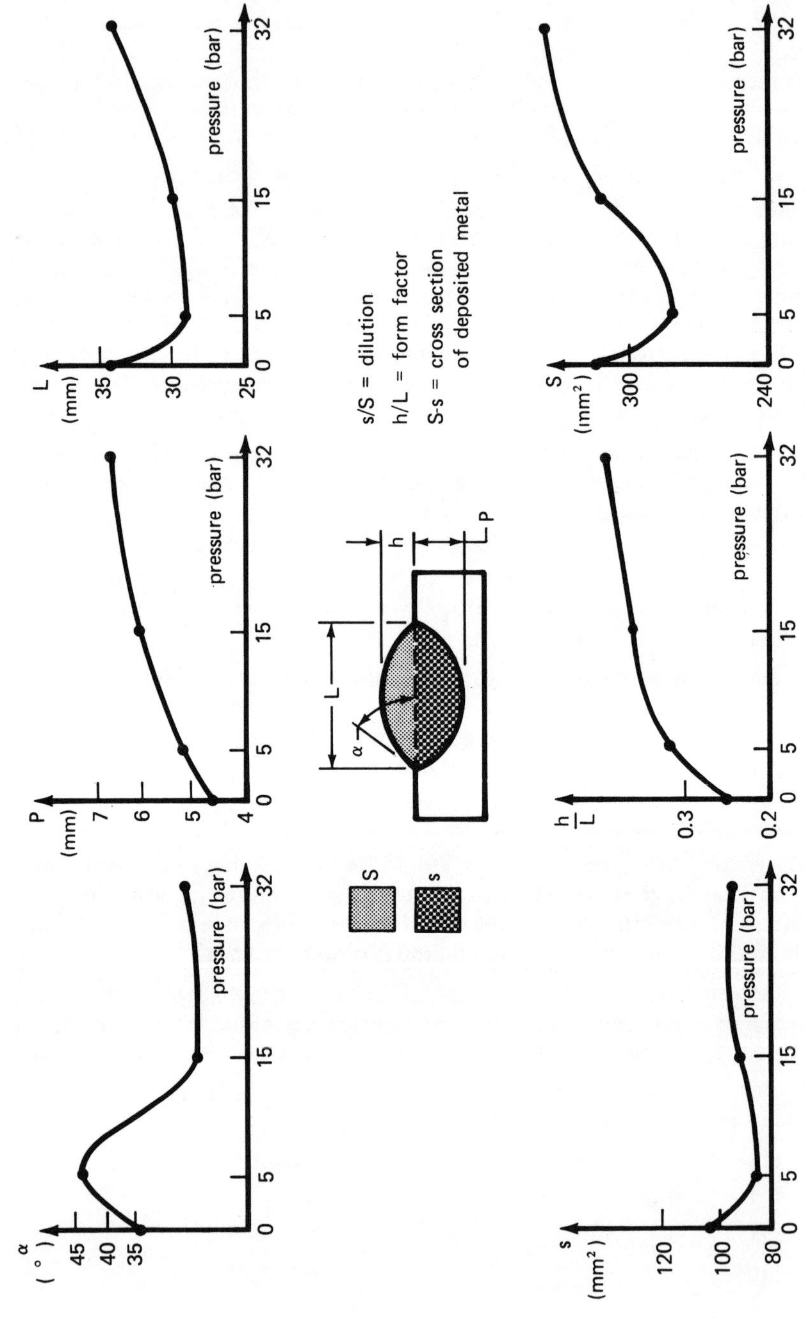

Fig. 5 – Geometry of weld bead in flat position

By recording arc voltage and amperage for each of these during the deposition of a bead in the flat position, it was possible to classify these power sources, especially in terms of regulating the arc length. The best results were obtained with the most recent type of power source.

Simultaneously with these recordings, it is possible to measure the maximum arc length permissible at various pressures. These results prove that in hyperbaric welding an adequate selection of the power source is of utmost importance for the quality of the weld (Fig. 6).

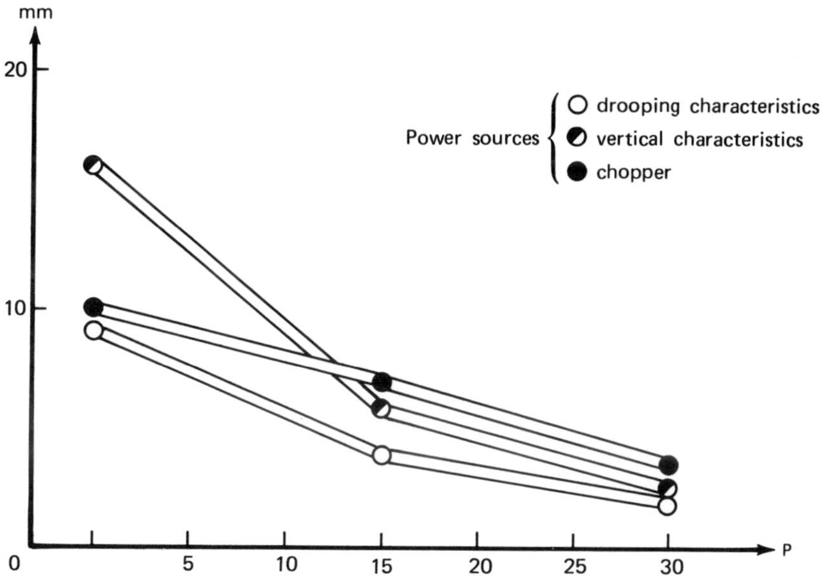

Fig. 6 – Maximum arc length vs pressure

Gas tungsten arc and gas metal arc welding. In gas tungsten arc (GTA) welding, the problem of the influence of pressure on operating conditions is different. Pressure acts directly on the constriction of the arc and the cathode spot is reduced, thus altering the arc stability; the arc tends to become erratic. Recordings of arc voltage (Fig. 7) show the variations in arc length resulting from this instability.

If we add the possible formation of local magnetic fields likely to cause true deflection of the arc, and also that the fluidity of the weld pool increases with pressure as does the difficulty of starting the arc, it becomes possible to appreciate even better why it is necessary to master all these factors to obtain highly reliable results.

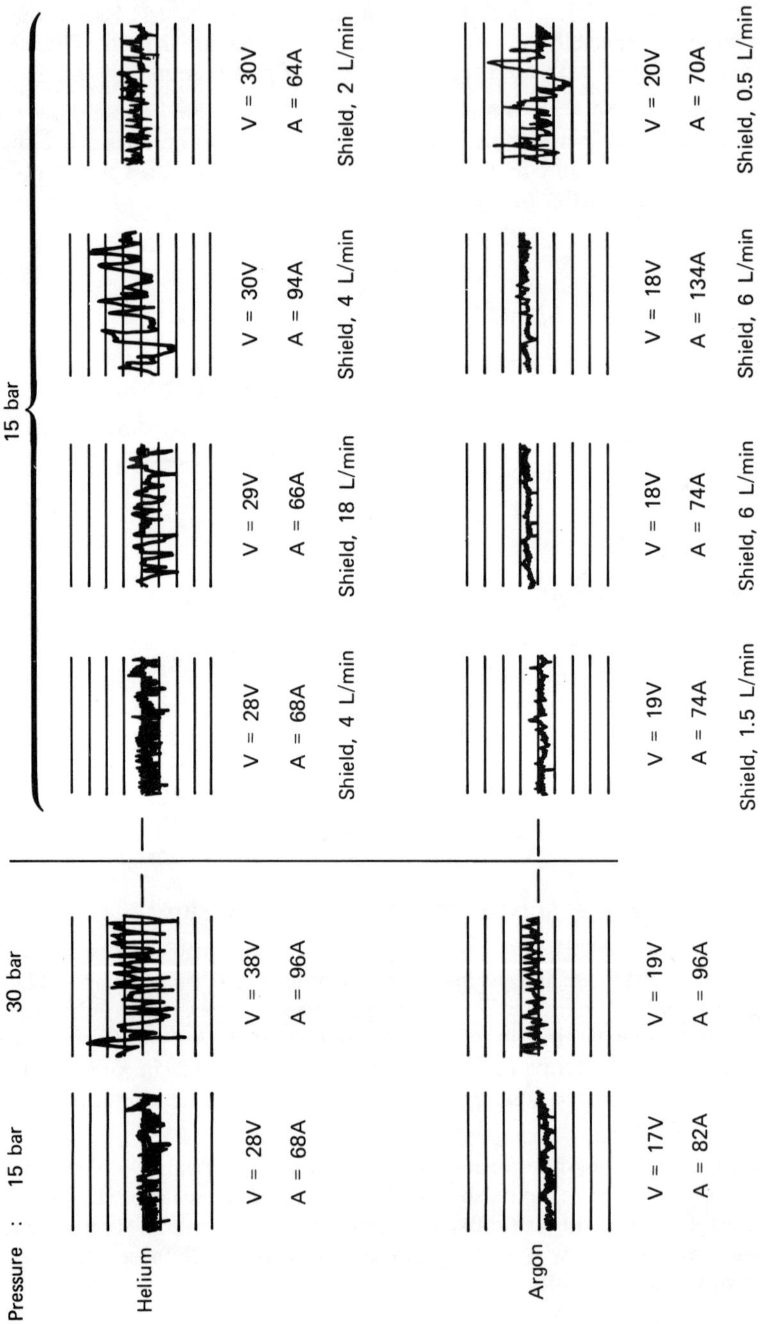

Fig. 7 – Arc instability in GTAW in terms of voltage

In gas metal arc (GMA) welding, the metal transfer across the arc is affected by pressure. This is caused essentially by a modification of the distribution of energy losses both in the electrode extension and in the arc. Therefore, conventional power sources with flat characteristics seem to be poorly adapted to the perfect control of energy within the arc during the short circuit period. Significant developments should be made by using microprocessors capable of controlling at any time the variables characterizing metal transfer in the arc, i.e., amperage, voltage, arc power, etc.

Metallurgical factors. Although pressure is too low to affect the actual structure of the base metal, this is true neither for the chemical reactions determining the composition of the deposited metal nor for the weldability of the material to be welded under hyperbaric conditions.

Chemical composition and mechanical properties. One of the most significant points is the variation in chemical composition of the metal deposited by iron powder, low hydrogen electrodes. This is due to the decomposition of the carbonates contained in the covering, leading both to an enrichment in carbon and oxygen and to a reduction in oxidizable elements such as manganese, silicon, etc. (Fig. 8). This results in a shift of both the ductile/brittle transition temperature, the impact strength, and, more generally, in a loss of toughness of the weld (Fig. 9).

Weldability. The weldability of a material is determined by its chemical composition, its hardenability, and the welding process associated with the thermal cycle that the base metal may undergo during the formation of the weld metal zone.

After trying to thus characterize weldability, let us say that it can be determined for a given material only by a progressive approach involving several tests. For instance, analyzing the thermal cycle of shielded metal arc welding evidences a 20 percent increase in the cooling rate between 800 and 500° C (1470 and 930° F) (Fig. 10). Depending on the base metal, this may produce a brittle structure in the heat-affected zone or higher underbead hardness, or both. Dangers of cracking are further increased by the fact that the solubility of hydrogen in weld metal increases with pressure (Fig. 11). Implant tests prove this by evidencing that the critical cracking stress decreases as pressure rises (Fig. 12).

Qualification of Welders and Procedures

If we consider the codes now in force concerning the welding of underwater structures, drafting welding procedures requires a thorough knowledge of all variables influencing the execution of a weld. All these codes include lists defining ranges of validity beyond which the procedure must necessarily be requalified.

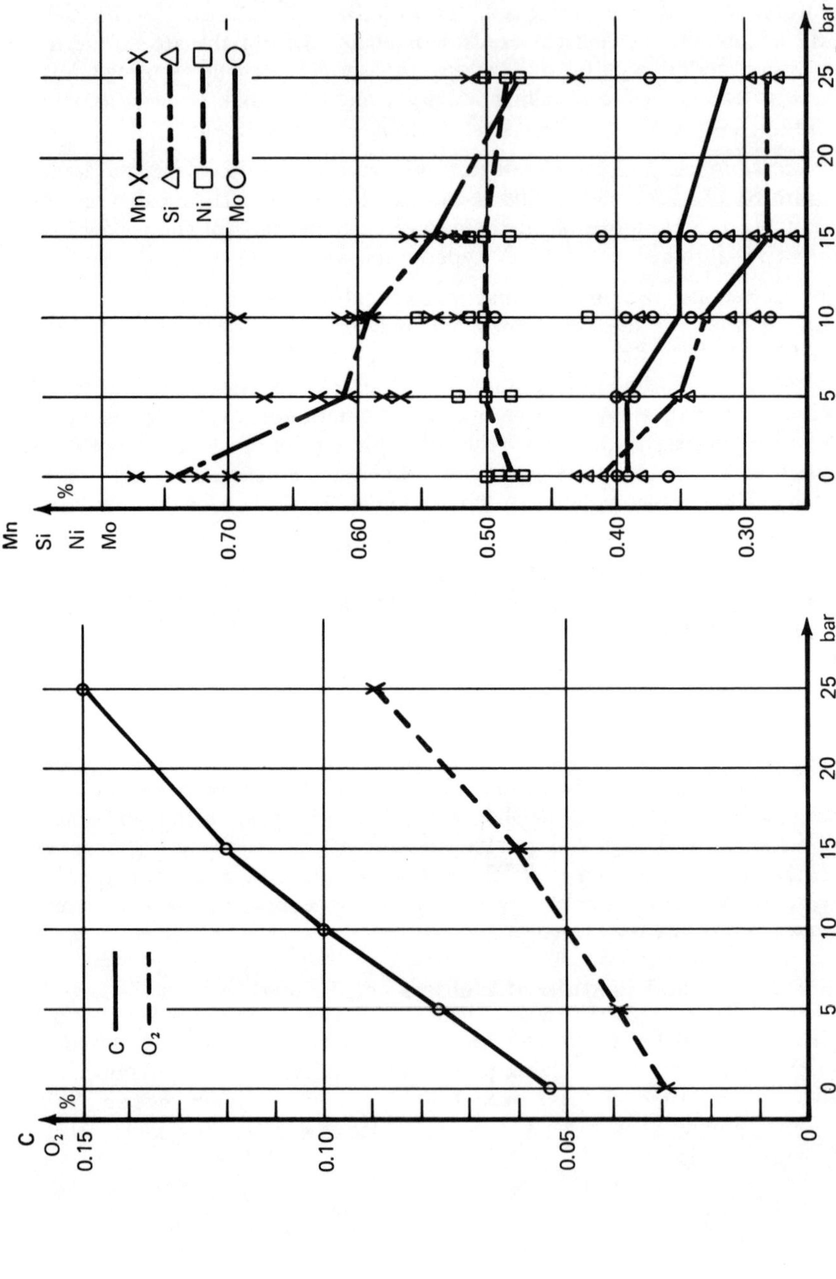

Fig. 8 — Effect of pressure on weld metal composition

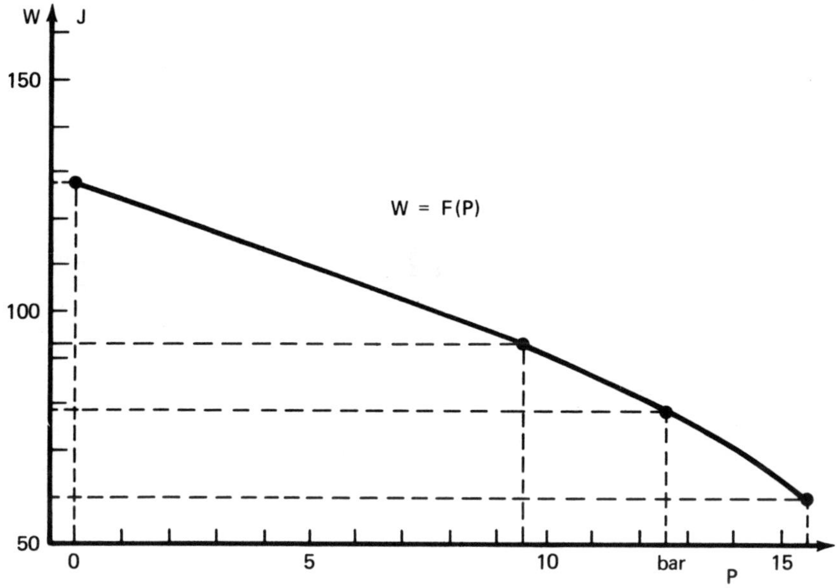

Fig. 9 – Impact toughness vs pressure

In fact, the few examples given in the previous chapter show that it is not possible to use new procedures without having previously examined how the new conditions introduced for these procedures affect the operating variables or the metallurgical factors as defined above. The most common example in this field is the application of a steel of a different origin than that of the steels qualified in the original procedure. Therefore, in view of the great number of steelmaking processes and of the coming in of grades such as X70 and X100 (API 5LX), a purely metallurgical approach of procedures will be difficult without a previous investigation of the effects of pressure on the weldability of these steels.

The same applies to tentative extensions of procedures for different depths, which require that a possible change in behavior of the covering be taken into account, thus modifying both the melting characteristics and the usability of the electrode.

Only tests made in the same conditions of pressure, temperature, humidity, etc. allow both new operating conditions and the validity of the procedure thus intended to be determined.

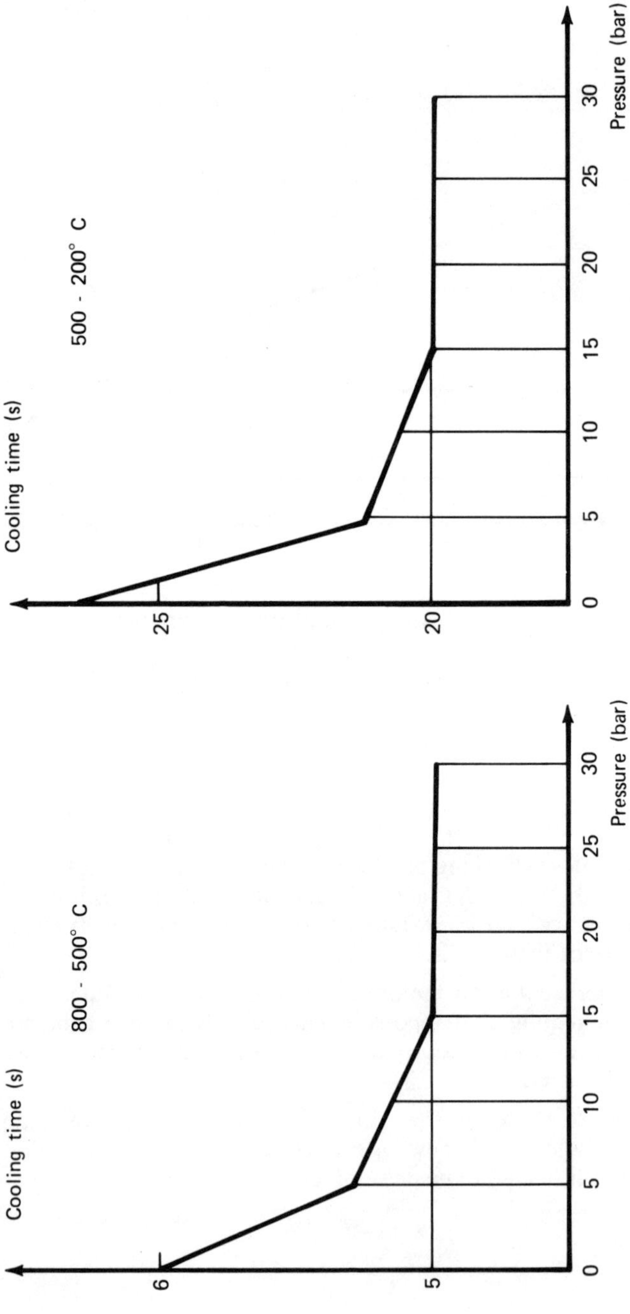

Fig. 10 — Cooling time vs pressure in HAZ (energy input, 14 kJ/cm)

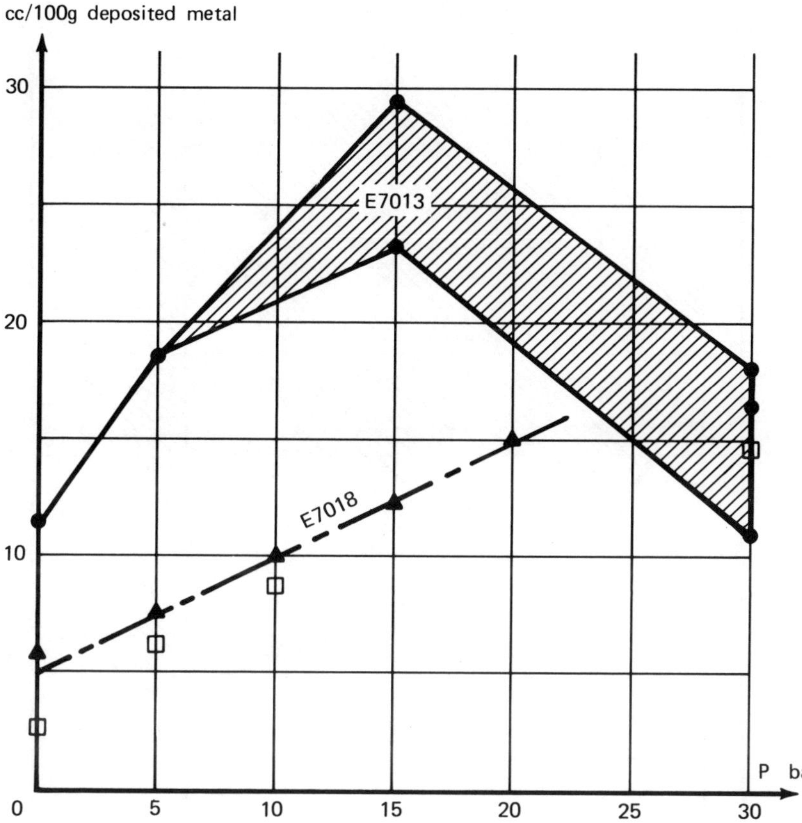

Fig. 11 – Diffusible hydrogen vs pressure (E7013 and E7018)

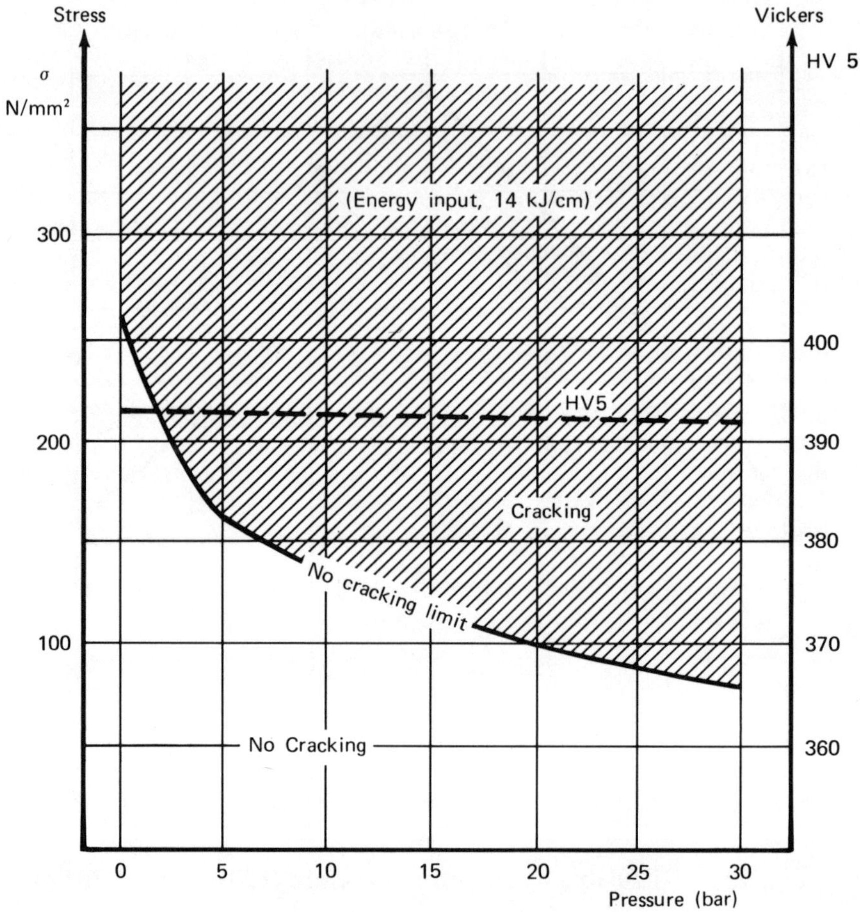

Fig. 12 – No cracking limit vs pressure

Use of a Sophisticated Simulation Chamber to Carry Out Preliminary Tests with a View to Developing Procedures

Before dealing with the use of a simulation chamber to carry out preliminary tests with a view to the development of procedures, let us ask whether it is justified to build such a facility. Regardless of the costs involved, two positions may be taken, depending on the attitude of the diving and oil companies: either they take a static position of preferring to use procedures that are tried and proven, not wanting to innovate; or they choose to master the problem and to innovate in order to develop increasingly better procedures that produce top quality results.

Of course, in the first case, building a chamber is not justified, but sooner or later, problems due to the technique not being mastered must be faced. In fact, it should not be overlooked that pressure is a new variable for the welder and only very few specialists are now capable of predicting certain anomalies encountered in hyperbaric welding. To a certain extent, this attitude may lead to a lack of safety in welding. For instance, during the first tests required to develop the procedure for the FRIGG pipeline in the North Sea, we were very surprised to find special types of defects caused by the type of covering and the chemical composition of the weld metal that no laboratory had ever evidenced before.

In the second case, the necessity for building such a facility is obvious. No matter how important this motivation is, the costs of tests aiming to improve, master, or develop a technique should always be borne in mind. For instance, because of the imperatives of diving, a manned full-scale test at a depth of 150 m for one day requires that the hyperbaric facility be mobilized for a minimum of 4 days, thus resulting in an expense of $60,000 to $100,000; whereas, for the same program, significant tests made in a simulation chamber would cost less than $15,000. In addition to this, simulation equipment enables risk taking with regard to the selection of certain welding techniques without endangering lives.

If such a facility is to meet the chosen objective, it must comply with a certain number of requirements, the most important of which are shown on Fig. 13.

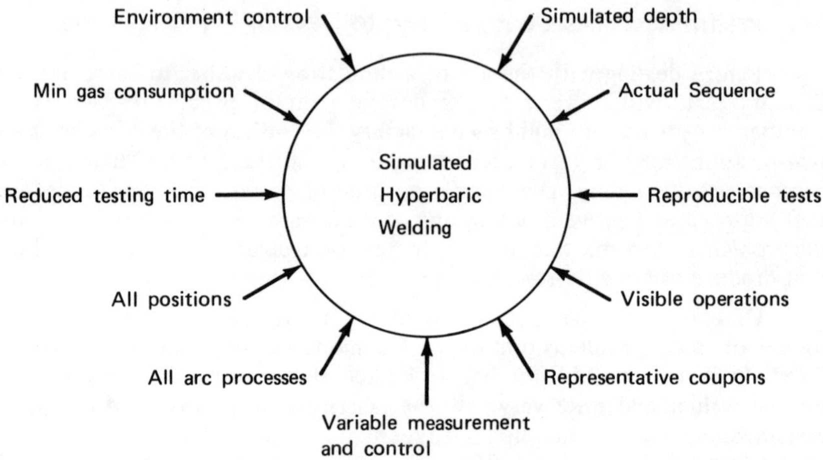

Fig. 13 – Objectives to be reached to simulate welding under hyperbaric conditions

On the basis of all these rules, the Institut de Soudure, i.e., the French Welding Institute, with the financial help of and in cooperation with TOTAL-CFP, has built a hyperbaric simulation facility called HYPLAB 3,300, which includes the following:

(1) A welding chamber with a capacity of 1,200 liters (42.35 ft^3), permitting the simulation of hyperbaric welding at depths to 1,000 m (3,300 ft)

(2) A unit for ambient gas processing and conditioning having the following characteristics:
   (a) Temperature: 20-40° C, ±3° C
   (b) Humidity: 40-100%
   (c) Removal: fumes, CO, CO_2
   (d) Oxygen content control

(3) A storage and compression unit for ambient gas

(4) A robot permitting the programmed simulation of a manual or automatic welding procedure, the various programs stored so as to ensure a good reproducibility of the welding operation

(5) A measuring and control unit that may be connected to a computer terminal for data acquisition and processing

The simulation facility permits:

(1) the use of all arc welding processes

(2) welding in all positions, 400 x 300 mm flat workpieces or 10 in. pipes, up to a maximum thickness of 40 mm

(3) the production of welded assemblies by following the actual welding sequences of the manual or automatic processes with a view to prequalification tests

(4) pre- and postweld heat treating

(5) welding in perfectly controlled and measured environments

By using this facility, it is possible to develop procedures and to investigate new automatic welding methods in all positions at depths to 1,000 m (3,300 ft). At present, the tests carried out at a simulated depth of 500 m (1,650 ft) allow the anticipation of an interesting future for automatic hyperbaric welding at depths yet unexplored by man.

Conclusion

The ever increasing energy demands of our civilization have led oil companies to prospect for and exploit submarine oil and gas deposits in areas of more and more difficult access and to build subsea pipeline systems to collect the gas or oil. Hyperbaric welding permits these pipelines to be perfectly welded on the seabed. However, the introduction of a new condition such as pressure modifies many factors governing the physical, chemical, metallurgical, and other phenomena specific to welding.

Building special test equipment is necessary to be able to reproduce the particular conditions of hyperbaric welding; these facilities must allow conducting tests under actual conditions of welding up to the stage of prequalification of a welding procedure.

The Consequences of Failure — The Products Liability Approach as It Pertains to Offshore Design Construction and Repair Work

John W. Robinson
Attorney, Lafayette, LA

The legal ramifications of failure as a result of the defective design, construction, maintenance, and repair of offshore platforms and pipelines is far-reaching and necessarily involves such diverse areas of the law as contracts, torts, and property law. Within the framework of these legal causes of action has developed a concept called "products liability," which is simply a term used to describe a type of claim for personal injuries or property damage arising out of the utilization of a product or perhaps even a service.

The concept of products liability is significant in that it provides a means of legal recovery by a third party who is not in privity with and/or has a contractural relationship with the party causing damage. This topic necessarily omits direct discussion of the written contractural obligations and liabilities that exist between the contracting parties to a construction and/or repair project.

Historically, privity of contract (direct contractural relationship) has posed the greatest problem in the expansion of the concept of products liability. In the *Winterbottom* decision, decided in 1842 by an English court, an individual who had contracted with the Postmaster-General to construct and repair mail coaches was sued by a driver as a result of an overturning coach. In denying recovery the court stated:

> "There is no privity of contract between these parties; and if the plaintiff can sue, every passenger, or even any person passing along the road, who was injured by the upsetting of the coach, might bring a similar action."

It is interesting to note that the only wrong committed by the coach contractor was in failing to bolt all four wheels of the mail coach securely to their axles.

With the coming of the Industrial Revolution, social policy gradually changed, and the courts began to recognize and adopt exceptions to the rule of nonliability of persons not in privity with a plaintiff, i.e., manufacturers of drugs, firearms, explosives that were considered inherently dangerous, etc.

The landmark decision of *MacPherson v. Buick Motor Co.,* 111 NE 1050 (1916), sealed the rejection of the privity rule in cases based upon *negligence* where the product involved could be regarded as dangerous, if negligently made. In *MacPherson,* one of the automobile wheels was made of defective wood, and its spokes crumbled into fragments. Even though the wheel had not been manufactured by Buick, there was evidence that the defect could have been discovered by reasonable inspection, which had not been done.

MacPherson is now recognized by all American courts, and its principles expanded by later jurisprudence to include liability by manufacturers to third parties for negligence in mislabeling or in failing to warn, for negligence in installation or inspection, or for an unsafe or defective design. In addition, the restrictive concept of the "inherently dangerous product" has all but disappeared from the products liability cases based upon negligence.

Although *MacPherson* established that a manufacturer did have a duty to the ultimate user of a product, in the absence of contract, that duty was only one of "reasonable care" under negligence principles. In an attempt to place a higher burden on the manufacturer of products, courts began to return to the old concept of "warranty" to provide a means of recovery in damage claims. In these claims, claims were based upon breach of promise, express or implied (not directly represented or written), to furnish products of fit and merchantable quality, without the necessity of proving negligence. Since this action was theoretically based upon contract, or sales law, the question of privity became a problem as regards claims by third parties not directly involved in the sales contract. However, by the early 1960's, many courts had finally eroded away the privity requirement in warranty cases, as they had done earlier in cases involving negligence.

Perhaps no doctrine in the law has had as much impact on our present society as that of the "Doctrine of Strict Tort Liability." It has been commonly referred to as "liability without fault." What it correctly means is liability without negligence. In these cases, the damaged party is generally relieved of the burden of proving that a defendant failed to act in a reasonably prudent manner in the design, construction, or sale of a product. Strict tort liability is not a new concept, but instead is a new cause of action based upon a melding of negligence and warranty principles. Louisiana is an exception to this rule since its civil code has always provided for specific instances of strict liability, such as in the case of damage caused by the ruin of a building.

The groundwork for strict liability was laid by many legal writers who saw the need for a simple, direct action by an injured user of a defective product against any and all of the persons responsible for its presence in the market place; an action unfettered by the technicalities of sales law and unburdened by the necessity of showing negligence in the manufacture or the construction of a product. Since strict liability is a tort concept, privity of contract has not presented an obstacle to recovery by third parties. By the late 1960's, strict tort liability had been recognized by a majority of states, either expressly or in principle, and has been codified in the Restatement of Torts, Second.

Lawsuits against the manufacturers of chattels or movables represented by and large the great majority of products liability litigation. The implications of this type of liability as it relates to the offshore oil industry is enormous not only because of the very technical nature of this enterprise, but because of the vast sums of money involved in its operation. Therefore, the consequences of failure offshore in terms of potential damage is staggering.

In the recent case of *Bradco Oil and Gas Co. v. Youngstown Sheet and Tube Co.,* 532 F.2d 501 (5th Cir., 1976), an oil company that has purchased high-strength tubing for use in drilling an exploratory oil and gas well sued the manufacturer for damages in the amount of $1,500,000.00 for the loss of the well as a result of fractured tubing at a depth of 1,500 feet. Even though the oil company had selected the type of tubing desired and provided the specifications, the evidence revealed that the tubing chosen was susceptible to embrittlement as a result of contact with a hydrogen sulfide environment. The question in this case related to whether the manufacturer had the duty to warn the oil company that such tubing could be dangerous when used in wells containing trace quantities of H_2S. This court, in recognizing the applicability of strict liability against manufacturers in Louisiana, held for the manufacturer in finding that the product possessed no inherent defect and that the defendant-manufacturer was not required to warn the oil company, a sophisticated purchaser, of dangers generally known in the industry to protect against use under adverse well conditions of which Bradco was aware, or should have been aware.

In the very interesting case of *Guilyot v. Del-Gulf Supply, Inc.,* 362 So.2d 816 (1978), a manufacturer was sued for personal injuries as a consequence of an alleged improperly designed pipe hook installed on a pipe barge. In this case, plaintiff was seriously injured when a 7000-pound pipe rolled into him. The court, in finding no fault in the design of the hook, reasoned that ease of removal of the pipe hook served the legitimate purpose of avoiding risks inherent in freeing a pipe hook with a retaining device and that the risks permitted were not unreasonably disproportionate to the risks avoided.

In applying the doctrine of strict liability to a manufacturer of a defective product under Texas law, the court in *Klein v. Continental Emsco,* 445 F.2d 629 (5th Cir., 1971), held that the user or consumer injured by a defective product must prove that such product was defective when it left the possession of the manufacturer. In this case a drilling rig was damaged when a traveling block struck the crown assembly and fell to the floor, dislodging a section of the rig. It was alleged that the damage was caused by a defect in the main air valve manufactured by Westinghouse. In the instant case, the court found that Westinghouse, whose valves were meddled with by Continental-Emsco in the creation of the rig console, could not be held liable solely upon the basis of a doctrine aimed at the fabricator of an integrated unit. Since the jury in the lower court had found no liability on the part of Continental-Emsco, and no appeal was taken, there was no further discussion of the issue of fabricator liability by this court.

The extreme in products liability litigation is seen in *Nobles v. Lincoln,* U.S. District Court, E.D. Tex., No. B-77-286 (1978), in which a jury awarded $500,000.00 for the wrongful death of a 56-year-old welder against five manufacturers of welding rods used by the decedent during his working life. The complaint alleged that the long-term use of the welding rods caused lower motor neuron disease and chronic bronchitis, which combined to cause death. Liability was based upon failure to warn the worker of the hazards involved.

A somewhat different approach to products liability litigation can be found in *Champagne v. Chevron U.S.A., Inc.,* 605 Fed. 934 (5th Cir. 1979). This case involved an action by an employee of an independent welding company that had been hired by Chevron to do welding work on a fixed offshore drilling platform. The employee sustained injuries when a nut within the nozzle of a fire hose came loose during testing operations and a sudden burst of water propelled the employee forcefully against a guard rail. Subsequently, the employee brought a diversity action against both Chevron and the manufacturer of the hose. After trial of the matter, the jury returned a $250,000.00 verdict against Chevron and also found that the nozzle had not been defectively manufactured. In affirming the verdict, the Court of Appeals cited the earlier case of *Olsen v. Shell Oil Co.,* which held that the owner of a building under the Louisiana Civil Code has a non-delegable duty to keep his building and all appurtenances thereto in repair and that an owner is "strictly liable" for failure to perform this duty. This court, in analogizing a fixed platform to a building, held that a fire hose is no less an appurtenance than a window fan, a water heater, or electrical wiring.

While the manufacturer of component parts or products is included within the sphere of products liability, so is the ultimate assembler or fabricator of parts and products. This is illustrated in the case of *Wood v. Kane Boiler Works,* 238 SW 2d 172 (1951), in which a wrongful death action was brought against the fabricator of steel pipe that burst during hydrostatic testing by an independent company hired by the purchaser of the pipe. In this case, the plaintiff had discovered a defect in a weld and, at his order, the pipe had been chipped out and rewelded. Thereafter, during testing, the pipe burst and plaintiff was killed. Rejecting the theory that the plaintiff had assumed the risk of the injury, this court found that there was a way prescribed by the industry which, if observed by the defendant, would have ensured that the seam of the fabricated pipe would be so fused by welding along its center that the pipe would withstand the test to which it was being subjected at the time it burst, and that the fabricator knew that such a test was to be conducted, yet failed to so weld as to avoid the defect.

Fabricators and manufacturers have often alleged as a defense to products liability litigation that extended use of a product should necessarily shield them from liability. This defense was rejected in *International Derrick and Equipment Company v. R.L. Croix,* 241 F.2d 216 (5th Cir., 1957), in which an employee of an independent contractor brought an action against the fabricator of an oil derrick for injuries sustained when a gin pole leg pulled loose at a weld and was jerked downward by the weight of a line and struck the plaintiff, who was working near the top of the derrick. There was evidence in this case to suggest that the defect in the weld could not have been ascertained prior to the accident because the welded parts had been galvanized. The court in finding against the manufacturer held, "...The lapse of seven years (from original construction) does not, per se, relieve a derrick manufacturer from liability for injuries caused by a defective weld overcoated with a galvanizer."

This same rationale was followed by the court in *Pryor v. Lee C. Moore Corporation,* 262 F.2d 673 (5th Cir., 1959), a case in which a derrick of an oil well drilling rig collapsed due to a defective weld at the foot of one leg of the derrick during a pipe loosening and pulling operation, causing injury to a rig employee. In this case, the appeals court remanded the case for a new trial on the basis that a defendant-manufacturer could still be held liable after a period of 15 years of rig use.

In the recent case of *Ramos v. Liberty Mutual Insurance Company,* 615 F.2d 334 (5th Cir., 1980), the Fifth Circuit Court of Appeals reversed a verdict in favor of a fabricator of an offshore drilling rig mast that had collapsed, injuring the plaintiff. Plaintiff alleged that the 3-part mast telescoped within itself when a pin that connected the segments failed. The appeals court held that it was error on the part of the trial judge in excluding evidence that 10 days after plaintiff's accident defendant delivered a new mast to plaintiff's employer. The new mast incorporated design features that strengthened the pins, which design innovations were in existence at the time of the accident.

One of the most difficult problems in evaluating products liability litigation is differentiating the standards of duty as regards manufacturers or fabricators and contractors. While a contractor's duties are ordinarily fixed by the terms of his contract, as a general rule there is implied in every contract for work or services a duty to perform them skillfully, carefully, diligently, and in a workmanlike manner, and a negligent failure to observe any of these conditions is a tort, as well as a breach of contract. Traditionally, it has been declared to be a rule of law that no cause of action in tort can arise from a breach of duty existing by virtue of contract unless there is between the defendant and the person injured "privity of contract." That is, a plaintiff in an action for negligence, who bases his suit upon the theory of duty owed to him by the defendant as a result of a contract, must be a party to or privy to the contract; otherwise, he fails to show that a duty was owed toward himself or that any wrong was done to himself.

While in the past, recovery by third parties injured by negligent performance of contractural duties has generally failed for lack of privity, this restriction in cases involving personal injury and economic loss is being slowly eliminated by the courts. The old rule of nonliability had particular application in the situation where one not a party to a construction contract attempted to recover for injuries received after completion of the project. Nonliability in these cases was based upon public policy to avoid endless litigation and an excessive burden on contractors. However, as time passed, courts began to make exceptions to the rule of nonliability of contractors, as in the case of a contract to be used for a particular purpose requiring security for the protection of life, such as scaffolding, or where the contractor is charged not only with the duty of installation, but also with the duty of continuing inspection. Likewise, an exception was made to nonliability in case of conditions produced by the construction, or articles constructed or installed that are inherently dangerous.

More recent cases generally have done away with the approach of nonliability with its many exceptions and have established the rule that the contractor is liable for injuries or damages to third parties after acceptance of the completed work by the contractee, where the work is reasonably certain to endanger third parties if negligently prepared or constructed. This view is based on the recognition by some courts that there are no sufficient grounds for a differentiation between the liability for negligence of a manufacturer of goods or products and a construction contractor.

A construction contractor who properly follows a set of plans and design specifications furnished by another, which are not obviously defective, generally will not be held liable as a result of damages sustained by a third party as a consequence of such construction. However, if the contractor itself makes design changes or innovations, then perhaps liability will attach. In the case of *Smith v. Ortloff Co.*, U.S. District Court, S.D. Ala., No. 75,301 (1977), a construction company that had entirely designed and built an oil refinery was sued by an employee of the owner of the refinery for damages as a result of sustaining brain damage after inhaling hydrogen sulfide at the plant. Because of certain innovations made by defendant in the design of the plant, a stream of cold liquid at 27° F was dumped into slop tanks containing liquid at 90° F. Plaintiff alleged that this caused a chemical reaction to occur, resulting in the release of hydrogen sulfide in dangerous quantities through an atmospheric vent on the top of the slop tanks. Plaintiff was overcome by gas while working near the tanks on the fourth day after the plant had opened. Defendant ultimately settled for $1,125,000.00.

In *Beach v. Anaconda,* U.S. District Court, E.D. La. (#73-685, 1978), the surviving widow and her two children brought a wrongful death action for the death of their husband and father when an ammonia loading line ruptured, releasing ammonia vapors that caused the instantaneous death of the decedent. Suit was brought against the executive officers of the chemical plant, the manufacturer of the flex hose that ruptured, the supplier of the component parts to the manufacturer, and the contractor who installed the hose at the plant. (One of the component part suppliers was dismissed on summary judgment.) The case was settled for $341,000.00

Even today, "privity of contract" still presents problems to courts dealing with construction contracts. In *Avondale Shipyard, Inc. v. Cuffe,* 434 F.Supp. 920 (1977), the court refused to recognize an action for indemnity based upon a breach of an implied contractural term of workmanlike performance, even though the United States Supreme Court so recognized such a cause of action in *Ryan Company v. Pan-Atlantic Steamship Co.,* 76 S.Ct. 231, a case involving maritime indemnity. In Louisiana, it appears that even today recovery can only be obtained by a third party against a contractor upon a cause of action based in tort upon negligence rather than warranty.

The *Avondale* case is particularly significant because of an attempt to hold an architect and design participant liable under a theory of strict products liability (liability without fault). The court, in refusing to accept this theory, held that neither constituted a "manufacturer" of the allegedly defective vessel. This statement by the court seems to represent the prevailing view nationwide that the doctrine of strict products liability is not applicable to providers of services such as independent engineers, architects, or even repairers (onshore or offshore). This was the holding in *Swenson Trucking v. Truckweld Equipment Co.,* 604 P2d 1113 (Alaska 1980), wherein the defendant repairer was sued for not discovering a faulty weld in the original manufacture of a truck's ram assembly. The court observed that strict liability generally applied to products, not services. The plaintiff was, however, entitled to maintain a negligence action for any negligent repair work performed by the defendant.

The case of *Block v. Fitts,* 274 So.2d 811 (1973), properly sets out the standard of care required of a repairer under Louisiana law (auto repairman). It held that a repairman who undertakes to repair a dangerous defect in a product, but who fails to do so, owes a duty to third persons, as well as to the owner, to advise the latter in some manner that the defect has not been repaired. If he represents to the owner that he has repaired the defect and that the product is safe, and he allows the owner to utilize the product under that erroneous impression, then the repairman's negligence in failing to properly repair the product, coupled with the failure to warn the owner, may be a proximate cause of an accident attributable to that defect.

While products liability litigation involving actions by third parties based upon defects in design, construction, fabrication, and repair of fixed offshore platforms and pipelines is relatively limited at this time, certainly from the current jurisprudence we can see a trend toward greater responsibility on the part of anyone who either sells, designs, manufactures, constructs, services, or owns a product or structure that is or may become defective and cause damage.

So strong has the trend been in products liability litigation that legislatures across the country have given serious consideration to proposals that would substantially alter product liability law, particularly in the area of strict liability, i.e., state-of-the-art defense based upon feasibility or knowledge has been offered.

The Management and Planning of Underwater Certification and Inspection

Peter Thornton
Tokola Underwater Engineers Limited

Introduction

No legal requirements are expressly stated for the maintenance of platforms in U.K. waters and survey requirements are very few. Repair and maintenance requirements are implied, but it is the requirements for regular survey that are briefly stated in The Offshore Installations (Construction and Survey) Regulations 1974.

The Regulations

The inquiry into the loss of the jack-up 'Sea Gem' in 1965 recommended that any mobile rig entering U.K. waters be surveyed and checked to be adequate for the North Sea environment. If periodic surveys revealed deterioration or damage, then the rig could be removed from site. The 1974 Act was extended to include fixed platforms, although many of the pre-1975 platforms do not comply with the Regulations, making long term certification difficult, so that special conditions have had to be imposed.

The Regulations require that on or after 31st August 1975, no platform shall be established or maintained in U.K. waters unless it has a valid certificate of fitness granted by an appointed Certifying Authority, of which there are six. A major survey must be carried out at intervals not exceeding five years. Additionally, an annual survey must be done to check on deterioration, but this can form a part of a continuous major survey. An additional survey may be required if damage or deterioration is suspected, which must be reported to both the Department of Energy and the Certifying Authority. Continuance of the certificate is dependent upon the platform remaining fit for its intended purpose.

Guidance Notes 1974

Subsequent to the issue of The Offshore Installations (Construction and Survey) Regulations 1974, the Department of Energy issued "Guidance on the Design and Construction of Offshore Installations," issued by H.M.S.O. 1974. This was intended as a guide to the technical standards required under the Act for the design and construction of an offshore structure and is an indication of the minimum standards that will be accepted when consideration is given to an application for a Certificate of Fitness. It does not purport to be fully comprehensive nor, unlike the Regulations, has it legal force.

Many of the recommendations made in the guidance notes are merely a duplication of what is contained in the Regulations.

The following extracts are all from the appendix of the document, which contains particulars of the general scope of the certification procedures (design assessment, surveys, etc.).

Major Survey

(1) Visual examination of the whole structure above and below water to assess its general condition to detect obvious damage, to determine the nature, extent, and thickness of marine growth and the condition of any protective coating; special attention should be given to areas of the splash zone.

(2) Detailed close visual inspection, after thorough cleaning of all nodes, joints, and welds (this may be reduced to not less than 10 percent of all nodes, etc.), if there is acceptable evidence of an equivalent survey within the preceeding five years. Where undertaken below water, these inspections are to be recorded by a method acceptable to the surveyor, e.g., video tape.

(3) Such additional close inspection or reinspections as the surveyor may judge necessary.

(4) In the light of (2) and (3) above, the surveyor may require further information, e.g., color photographs, stereo photographs, non-destructive tests, to verify the existence of and/or extent of suspected defects.

(5) Seabed level inspection for abrasion, damage, and local scour; also to examine the condition of skirts, aprons, and other long-term scour control devices.

(6) Cathodic protection potential survey and inspection of the seabed to detect the presence of metallic debris likely to affect the working of the system (the seabed survey may, of course, be combined with (1) or (5) above).

(7) Inspection of sanitary and overboard discharge lines and their values and other closing arrangements.

(8) Measurement of settlement and differential settlement.

After the first certificate of fitness has been issued, the subsequent survey cycle may be carried out on a continuous basis in accordance with the schedule. That should provide for the equivalent of a major survey and associated annual surveys within the period of the certificate.

Annual Surveys. The purpose of carrying out an annual survey on the structure is given as follows:

> "The purpose of the annual survey is to ensure that any deterioration of the structure is within acceptable limits; that secondary structures and fittings concerned with the safety of the installation and the safety of personnel are in sound condition."

The annual survey is thus intended as a check on the rate of deterioration of the structure in between major surveys.

The following schedule is given as an example of an annual survey:

(1) General visual inspections down to and including the splash zone to detect obvious damage and indicate areas likely to warrant further investigation.

(2) Close detailed and fully recorded inspections down to and including the splash zone, as the surveyor may select, subject to not less than 10 percent of the total number of joints, should be inspected and attention should be concentrated on areas and features where experience suggests that problems are likely to arise. The steel should be thoroughly cleaned to facilitate the detection of incipient defects.

(3) A close inspection of any underwater repair work irrespective of depth undertaken since the last survey and, if the manager of a fixed installation has had occasion to take emergency measures to control scour, the inspection should include the condition of any permanent scour prevention works. Following above inspections, the surveyor may require that nondestructive tests be carried out to verify or eliminate suspect defects.

(4) General inspection of secondary structures and fittings concerned with the safety of the installation and personnel.

(5) At fixed installations, measurement of settlement and differential settlement.

The second annual survey after each major survey should include a general inspection of major parts of the installation below the splash zone to determine whether any change had occurred in the condition of the installation since the main survey was carried out.

In light of experience gained from undertaking these surveys, the Department of Energy updated and revised the original Guidance Notes in 1977.

Offshore Installations: Guidance on Design and Construction, Second Edition, July 1977

This new document, hereafter known as the Second Edition, basically indicates that the Department of Energy's future approach concerning certification will be to put the onus on formulating an *adequate* survey program to meet the certification requirements on the installation owner and certifying authority.

The following is a summary of these new regulations:

(1) General Procedure. The owner is responsible for arranging surveys as and when they become due. In view of the problems of access, particularly below water, an agreed schedule of inspection and tests should be drawn up in each case; this schedule should be arranged by the owners and certifying authority . . . . . and should be sufficiently flexible to allow adjustments to be made in light of findings as the survey proceeds. The certifying authority should agree particulars of destructive and N.D.T. . . . . . . and the number and frequency of which tests should be made. Surveys are to include, where appropriate, the use and interpretation of any instruments installed to monitor structural behavior.

(2) Major Surveys - Recertification. The object of this survey is to ensure that any deterioration is within acceptable limits and that the installation continues to comply with all relevant requirements of the Regulations. The most pertinent comments covering surveys are:

(a) The initial inspection schedule should take account of the nature of the deterioration to which steel structures are liable . . . . . and of regions in which defects are most prone to occur . . . . . and of members or regions known to have been, or likely to have been, highly stressed or subjected to severe fatigue loading.

(b) Subject to the need to ensure the continued safety of the installation, the inspection schedule should require deep diving ONLY where no other satisfactory means exist of carrying out essential work.

(c) In a jacket type structure, the outer, more accessible members may reasonably be taken as representative of the internal members, unless original design or service history suggest otherwise.

(d) Service history and/or design assessments will suggest areas requiring special attention and, if no other defects are detected, other deep water areas need receive only spot checks.

It should be noted that the Certificate of Fitness does not apply to conductors, marine risers and tubes, since they are not considered as part of the structure. However, they impose loads on the structure and these loads should be taken into consideration in defining inspection areas.

(3) Annual Surveys. The purpose of the annual survey is to ensure that any deterioration of the structure is within acceptable limits, that secondary structures and fittings are in a sound condition. Annual surveys should include a close visual inspection down to and including the splash zone to detect obvious damage and indicate areas likely to warrant further investigation. In light of this inspection, the surveyor may request that further inspections or tests be made below water. The surveyor should cause a close examination to be made of any underwater repair work since the last survey.

Following the above survey, the inspector may require that N.D.T. be carried out. An assessment should be made of the thickness of marine growth on typical members or areas of the structure.

The second annual survey after each major survey should include a general inspection of major parts of the installation below the splash zone to determine whether any change has occurred in the condition of the installation since the main survey was carried out.

(4) Operations Manual. The Second Edition of the Department of Energy's Guidance Notes specifies that an Operations Manual shall be prepared for every installation for the use of the manager covering the strength, stability, and seaworthiness of the installation. This manual shall cover, among other things:

(a) The environmental conditions associated with normal working conditions
(b) The stage at which marine growth should be removed
(c) The characteristics of the installations foundations
(d) The use and interpretation of any instruments installed to monitor structural behavior, watertight integrity, foundation condition, and any other factors relevant to the continued safety of the installation

It is clear from this that in the future the platform operations manager must have access to data defining limiting levels of deterioration that are considered acceptable for the continued operation of the platform.

From the above requirements called for in the Guidance Notes, it can be seen that it is extremely important to have a well defined and cost effective inspection program prepared prior to the installation of the platform. Thus, the following criteria must be considered by the operator, to ensure that the necessary results are achieved with the least amount of delay and within an acceptable budget.

The requirements and problems of underwater maintenance must be understood by the designer at the inception of the platform's design. He must know which parts are to be inspected, and why. He must be advised what tasks can be undertaken underwater, what equipment can be used, and its effectiveness. He must be aware of the problems of diver access, the different types of vehicles that are on the market capable of undertaking the work, and the problems of navigation around the steel jackets or concrete structures; and he should incorporate a system of identification of major areas and nodes that will facilitate the inspection.

Access to a platform must be examined to see whether it is cost effective to build an inboard diving system or to build in a facility that would allow it to take an inboard system for the purposes of carrying out routine inspection over its lifespan; but if it is a concrete structure, access to the columns is limited because of the overhang. Another consideration is to build in a light-weight (or part saturation) diving facility to carry out inspection in the splash zone and down to the 50 m limit of air diving. Inspection items such as seawater intakes should, if possible, be placed within the air diving limit, below which the mode of operation and costs change appreciably. When inspecting vast concrete areas, location references are especially necessary so that return to some spot can be readily achieved.

(5) Fabrication Survey. The contractor nominated to carry out inspection should be involved not only during the design phase but during fabrication as well. Experience has shown that final fabrication can be completely different from drawings. A detailed visual survey of the structure should be carried out to photograph critical areas and critical nodes that might have to be inspected at a later date. A basic record is thus compiled for future inspection. Painted fabrication codes should be removed, as those can hinder divers, and be replaced by a system of identification relating to each node.

(6) Planning. An inspection program can now be prepared using analysis of the initial design and survey of the structure. A hierarchy of highly sensitive areas and areas subject to high stress can be produced, identifying when those areas should be inspected during the five year certification period. The subcontracted inspection companies can then be told exactly what has to be inspected, where it is, and why it has to be inspected. Formerly, organizations have put divers underwater and returned with lots of video tapes that do not show anything. A precise knowledge of everything that has to be done enables surface support facilities to be determined and the total inspection cost to be calculated.

Many surface support vessels are now being built that are designed purely for underwater inspection and maintenance. Operators should question the cost effectiveness of some of these vehicles and whether such a complex system is really needed; that can be assessed from the detailed inspection matrix. A vessel costs up to $40,000 per day, and it may be more cost effective to charter a vessel as required, rather than carry around large components, large diving systems, and heavy craneage unused. This reinforces the usefulness of having a facility on the platform to take a saturation diving system or some sort of system to implement repairs at a later date.

Conclusion

The management of underwater maintenance combines design engineering to ensure quality, the planning and scheduling of routine inspection, and the supervision of planned maintenance and repair with the contingency against the unexpected by insurance.